A Friendly Mathematics Competition

35 Years of Teamwork in Indiana

© 2003 by

The Mathematical Association of America (Incorporated)

Library of Congress Catalog Card Number 2002107971

ISBN 0-88385-808-8

Printed in the United States of America

Current Printing (last digit):
10 9 8 7 6 5 4 3 2 1

A Friendly Mathematics Competition

35 Years of Teamwork in Indiana

Edited by

Rick Gillman
Valparaiso University

Published and Distributed by
The Mathematical Association of America

MAA PROBLEM BOOKS SERIES

Problem Books is a series of the Mathematical Association of America consisting of collections of problems and solutions from annual mathematical competitions; compilations of problems (including unsolved problems) specific to particular branches of mathematics; books on the art and practice of problem solving, etc.

A Friendly Mathematics Competition: 35 Years of Teamwork in Indiana, edited by Rick Gillman

The Inquisitive Problem Solver, Paul Vaderlind, Richard K. Guy, and Loren L. Larson

Mathematical Olympiads 1998–1999: Problems and Solutions From Around the World, edited by Titu Andreescu and Zuming Feng

Mathematical Olympiads 1999–2000: Problems and Solutions From Around the World, edited by Titu Andreescu and Zuming Feng

The William Lowell Putnam Mathematical Competition 1985–2000: Problems, Solutions, and Commentary, Kiran S. Kedlaya, Bjorn Poonen, Ravi Vakil

USA and International Mathematical Olympiads 2000, edited by Titu Andreescu and Zuming Feng

USA and International Mathematical Olympiads 2001, edited by Titu Andreescu and Zuming Feng

MAA Service Center
P. O. Box 91112
Washington, DC 20090-1112
1-800-331-1622   fax: 1-301-206-9789
www.maa.org

121503 - 3515 m8

Preface

A college level mathematics competition has been held in Indiana for more than 35 years as of the writing of this book. Orginally known as the Indiana Small College Mathematics Competition, its official title became the Indiana College Mathematics Competition. However, many people refer to it by its popular nickname "The Friendly Exam" earned because of the collegial nature of the competition and the modest level of competitiveness between the participants and the institutions involved. It is really more about getting together to do mathematics and to challenge oneself than it is about winning a competition.

This book shares the efforts of the many students and faculty who have participated in the Indiana College Mathematics Competition over its first 35 years.

I would like to specifically identify the individual faculty members who have composed the exams. Unfortunately, this is the one piece of the history that has been lost over the years. Their attentive work at identifying or creating good questions has been a major reason why the competition has been a success for so long.

Over the years, exam authors have borrowed many classic problems from other sources. These have been included to provide a complete record of the exam, but have not generally been attributed to an original source.

I'd like to take this opportunity to thank all of the people who have made this book possible by helping in one way or another. Among all of these people, the following individuals provided noteworthy help: James Lee, Ken Luther, Steve Shonefeld, Pat Sullivan, Eric Tkaczyk, and the Purdue University Math Club.

I would also like to thank Pete Edson for having the idea for the competition, and extend a very special thank you to Paul Mielke on

behalf of myself and my colleagues in Indiana. Without Paul's efforts
to develop the contest, to support it over the years, and to lead the section
in building the endowment, Indiana and the MAA would not have this
wonderful example of a "friendly" competition that does so much to build
the mathematical community in Indiana.

Contents

The Indiana College Mathematics Competition: A Short History

Paul T. Mielke

On April 27, 1965, Peter Edson, a trustee of Wabash College, sent our dean a memorandum that included a newspaper clipping about a unique mathematics competition that was held among high schools in New Jersey. In this competition, each school entered a team that worked as a team on a set mathematics examination. Edson wondered if anything of that sort was done at the college level. I answered that I knew of no such competition but that I would be willing to suggest the idea to my Indiana colleagues. Thus, on May 18, 1965, I wrote a letter to heads of departments in the small colleges of Indiana relaying Edson's suggestion. Response to my letter was immediate and favorable, so I wrote a second letter, which read in part:

The Mathematics Department of Wabash College is pleased to invite its counterparts in the other colleges of Indiana to a friendly mathematics competition to be held on Saturday afternoon, March 12, 1966, at Wabash College, Crawfordsville. The purpose of this competition is to stimulate friendship and mathematical interest among members of the various departments and their undergraduate students. Program for the day, which we are calling "Pete Edson Day" in honor of the Wabash College trustee who conceived the idea for the contest, is as follows:

11:30 A.M. – 12:30 P.M. Registration, Baxter Hall Lounge, Wabash College
12:30 P.M. — Luncheon, South Mezzanine, Campus Center. Participants will be guests of the College.
2:00 P.M. – 4:00 P.M. — The competition, a written examination for undergraduates

1

2:00 P.M. – 4:00 P.M. — Informal seminar for members of the
various departments
4:00 P.M. — Coffee hour at which test papers will be gathered,
Baxter Lounge

The test, prepared by Professor George F. Springer of Indiana
University, will be similar to, though (one hopes) not as difficult as,
the Putnam test. There will be a significant difference, however. Each
school is invited to bring a team of three undergraduates who will
work on the test as a team, consulting with one another as needed.
Tests will be sent to Professor Springer for grading and announcement
of results. A suitable prize will be awarded by Wabash College to
members of the winning team. Ranking of other teams will not be
revealed by Professor Springer.

This first contest was held as scheduled, with eight schools partici-
pating: Anderson College, DePauw University, Earlham College, Franklin
College, Marian College, Rose Polytechnic Institute (now Rose-Hulman),
Valparaiso University, and Wabash College. In the spring of 1967 the
second competition was held at Marian College with 15 participants:
Anderson, Butler, DePauw, Earlham, Evansville, Franklin, Manchester,
Marian, Rose-Hulman, St. Joseph's, St. Mary-of-the-Woods, St. Mary's
(of Notre Dame), Taylor, Valparaiso, and Wabash.

Since its beginning, the contest has been called familiarly, "The Friendly
Math Competition", from the phrase used in the above letter. It has
been held every year, a total of 24 times as of 1989, with a minimum
participation of eight schools in that first year, a maximum participation of
nineteen in 1972, and an average participation of over thirteen schools. The
nineteen participants in the 1972 competition were Butler, DePauw, Earlham, Evansville, Franklin, Huntington, IUPUI Indianapolis, IU Northwest,
IU Southeast, Manchester, Marian, Oakland City, Purdue Calumet, Purdue
Ft. Wayne, Rose-Hulman, St. Mary-of-the-Woods, Taylor, Valparaiso, and
Wabash.

Throughout its history, the essential nature of the contest as a team
competition has remained the same, with a team consisting of at most
three members. The tests have usually been set and graded by a single
person, most often a professor from one of the large state universities.
Duration of the tests has been maintained at two hours. Before each test,
a team member from each school draws a number which is given to an
impartial judge, who attaches the number to the school's name in a memo.

Teams are instructed to identify their test papers with only their numbers, so that only these are seen by the person who sets and grades the test. Upon completing his/her work, the grader announces the results by team number to the judge who holds the identification key. This person in turn reports the results to the contestants.

Some other ground rules developed naturally. There was early agitation for publicly announcing the ranking of the first three teams. This idea was accepted, but it led to another problem. A school may enter any number of contestants in the Putnam competition, but it must name its team of three for the competition beforehand. Rose-Hulman has always been a worthy adversary in this Indiana competition, and a decision had to be made with regard to its wish to enter several teams. A consensus was reached that multiple teams of three could be entered, and that a school need not designate "its team." It was then decided that public announcement of the top three places in the competition would refer to schools rather than teams, so that if two teams from a given school scored second and third, for instance, the school would be awarded second place in the competition, and third place would be awarded to the next highest school.

Until 1987 the competition was restricted to the small colleges of Indiana, including branches of IU and Purdue, but in 1987 it was opened to all colleges, including the large universities.

The contests have always been managed by the host schools, though in its fall meeting of 1977, the Indiana Section of the Mathematical Association of America agreed to make them a part of its spring meeting, and this practice has been followed since 1978. Maintaining the "Edson Trophy" and awarding of prizes was assumed by Wabash College, which had a small fund for that purpose, to which Mr. Edson contributed while he was alive. Since this fund was exhausted, I have had the pleasure of supplying both trophy and prizes.

Each member of the winning team is awarded a suitably inscribed book prize, and the winners' names are inscribed on the Edson Trophy . The books are traditionally selected from those offered by the MAA. The Edson Trophy is a rectangular walnut plaque measuring 9×12 inches. To it is affixed a permanent identifying plate with this inscription:

THE
PETER EDSON TROPHY
INDIANA COLLEGE
MATHEMATICS COMPETITION

Below this plate is a smaller one containing the year and the name of the most recent winning college and its team members. Each year, this smaller plate is moved to the back of the plaque, and a new one replaces it. The trophy "travels"; it resides each year at the winning college. The back of the first trophy was filled by 1978, so it resides permanently at Wabash College, which was the winner in that year. The current trophy is at Purdue University, the 1989 winner. (Since writing the above, I have succeeded in making individual trophies for each of the past winners, affixing all the old winners' plates to them. The new trophies are made of sassafras, one of Indiana's most beautiful hardwoods. The two old trophies have been retired to Wabash College, where they record its six wins. Three special walnut trophies have been made to hold the 12 winning plates for Rose-Hulman. Henceforth, a non-travelling trophy will be awarded to each winner.)

It should finally be noted that the task here has been to write examinations that will challenge and separate the competitors, yet present some problems that all can solve, so that there will be enjoyment for everyone involved. As one might guess, the actual examinations varied in their success according to these criteria.

An Update of the History of the ICMC

I was introduced to the Indiana College Mathematics Competition (ICMC) during the spring of 1987, during my first year on the faculty at Valparaiso University. This was, as you read in Paul's comments, the first year that the competition was opened up to the large state universities and explains why I had not known of the competition while an undergraduate student at Ball State University.

I have been an enthusiastic supporter of the competition ever since. While the nickname "friendly" has not been used very much in recent years, the competition still maintains the same friendly approach that it had when it began. It is always very exciting and refreshing to be around the students from the various schools as they wait for the exam to begin, and to hear their conversations with each other and with their faculty advisors after the exam. In many years, a lot of friendly banter and good mathematical conversations occur among my VU students during the drive back to campus from the meeting.

In 1992, I became the Indiana section's first Student Chapter Coordinator and one of the responsibilities was to manage the ICMC. The first competition that I organized was at St. Mary's College in Notre Dame, Indiana, at a joint meeting of the Indiana, Michigan, and Illinois sections. Thirty teams from the three states participated in the competition that year. Subsequent Student Chapter Coordinators have helped implement changes to the ICMC that have made it the central part of the section's efforts to reach undergraduate students.

Since 1990, the average number of teams participating in the ICMC has been 25, generally from twelve or thirteen colleges or universities. Occasionally, a team from the Indiana Mathematics and Science Academy also participates. The largest competition was at Ball State University in

5

1998. That year, 29 teams from 22 institutions participated. Somewhat surprisingly, the large state universities have not dominated the competition since their inclusion, in either the number of teams participating or the number of competitions won.

The most significant change in the competition came in the form of a decision to move the competition from Saturday afternoon (on the second day of a day and a half meeting) to Friday afternoon as a lead off event for the meeting. Students can then stay for a banquet and after dinner speaker. Rather than having a single person grade the exams, a team of faculty graders do this work overnight. On Saturday, there is a session where solutions to the exam are discussed and the winners of the competition are announced late in the day during the section's business meeting.

The institution hosting the ICMC and the section meeting is invited to have one of its faculty members write the exam and solutions. If they decline this opportunity, the section finds an external source to do this work. Each team competing pays a $5.00 registration fee. Originally this fee was paid to the exam writer for composing and grading the exam. Now these fees are paid to the writer for writing the exam and for leading the problem solving session on Saturday.

This format has increased the number of students participating in the mathematics sessions of the meeting, both as presenters and as members of the audience. It has also established a new structure for the section meeting. Sessions on professional and pedagogical issues are now held on Friday afternoons during the competition itself, so that mathematical presentations are the central focus when students are part of the audience. The after-dinner speaker on Friday night is selected with the knowledge that at least half of the audience of 100–120 individuals are students. The section has also been able to experiment with late night workshops for students on Friday evenings, knowing that students will be staying overnight for the Saturday portion of the meeting.

By 1996, the Indiana section had established a small endowment to support the ICMC. This endowment will enable the section to continue to hold the exam each year and present prizes to the winning team members and the winning institution in the form of books and the Peter Edson Trophy, respectively. Proceeds from the sale of this volume will go the Indiana section to support undergraduate student activities in the section.

Rick Gillman
Spring, 2002

Exams

Exam #1–1966

As stated in the introduction, the first "friendly competition" was held at Wabash College, located in Crawfordsville, a bit northwest of Indianapolis. Eight schools participated in the competition that year. It was won by the team from Wabash College consisting of James Clynch, Albert Hart, Jr., and Larry Haugh.

P1966-1. Show that the equation $x^2 - y^2 = a^3$ always has integer solutions for x and y whenever a is a positive integer.

P1966-2. Consider any five points P_1, P_2, P_3, P_4, P_5 in the interior of a square of side length 1 (one). Denote by d_{ij} the distance between points P_i and P_j. Prove that at least one of the distances d_{ij} is less than $\sqrt{2}/2$.

P1966-3. Let p be a prime number and let $a_1, a_2, \ldots, a_p$ be integers not necessarily arranged in consecutive order and with possible repetitions. Establish the existence of integers m and n such that $1 \le m \le n \le p$ and such that $\sum_{j=m}^{n} a_j$ is divisible by p.

P1966-4. Two functions of x are differentiable and not identically equal to zero. Find an example of two such functions having the property that the derivative of their quotient is the quotient of their derivatives.

P1966-5. For two given positive integers n and k, how many different sequences of positive integers $a_1 \le a_2 \le a_3 \le \cdots \le a_k$ are there in which $a_k \le n$?

P1966-6. A sequence $\{x_n\}$ is defined by the following rule: $x_{n+1} = \sqrt{ax_n^2 + b}$ with $x_1 = c$. Show that this sequence converges whenever

$0 < a < 1$ and $b > 0$ regardless of the value of the real number c, and determine the limit of the sequence.

P1966-7. By an interval we shall mean a set of points x on the real line satisfying $a \leq x \leq b$ for a pair of real numbers a and b with $a < b$. Suppose that we are given a collection of intervals $I_1, I_2, \ldots, I_n$ which cover an interval I; that is $I \subseteq \bigcup_{k=1}^{n} I_k$. Prove that we can select mutually disjoint intervals from this collection which cover at least half of I.

P1966-8. Let us assume that a given pair of people either know each other or are strangers. If six people enter a room, show that there must be either three people who know each other pairwise or three people who are pairwise strangers.

Exam #2–1967

This competition was held at Marian College in Indianapolis. The winning team, consisting of David Hafling, Albert Hart Jr., and Robert Spear, was again from Wabash College.

P1967-1. $A = \{a_{ij}\}$ is a symmetric (i.e., $a_{ij} = a_{ji}$) $n \times n$ matrix with n odd, and each row of the matrix is a permutation of the integers $1, 2, \ldots, n$. Prove that the main diagonal is also a permutation of $1, 2, \ldots, n$.

P1967-2. Two parabolas have parallel axes. Prove that their common chord bisects their common tangent.

P1967-3. Show that for all $a \geq 0$ and $b \geq 1$, $ab \leq e^a + b(\ln b - 1)$.

P1967-4. For each positive integer n the binomial coefficients $\binom{n}{r}$, $0 \leq r \leq n$, are integers, some odd, some even. Show that for each n the number of odd binomial coefficients is a power of 2.

P1967-5. Show that if $|a_n| < 2$ for $1 \leq n \leq N$, then the equation $1 + a_1 z + a_2 z^2 + \cdots + a_N z^N = 0$ has no root z such that $|z| < 1/3$.

P1967-6. Prove that if the set S of points in or on the boundary of the unit square is partitioned into three disjoint sets A, B, and C, i.e., $S = A \cup B \cup C$ and $A \cap B = A \cap C = B \cap C = \emptyset$, then the least upper bound of the diameters of A, B, and C is greater than or equal to $\sqrt{65}/8$. The diameter of a set is the least upper bound of the distances between two points of the set.

P1967-7. Given $a > 0$ and $x_0 > 0$, show that there exists one and only one sequence of positive numbers $\{x_0, x_1, x_2, \ldots\}$ such that

$$x_n = \sum_{j=n+1}^{\infty} x_j^a,$$

for $n = 0, 1, 2, \ldots$.

P1967-8. Let T be a mapping of the Euclidean plane into itself which preserves all rational distances. Prove that T preserves all distances.

Exam #3–1968

Held at Franklin College, located just south of Indianapolis, this competition was won by a team from Earlham College. The team members were William Roha, Thom Sulanke, and William Wilson. It is unique because it is the only competition that came with a warning.

WARNING: The statements below should be viewed as conjectures. At least one cannot be done.

P1968-1. Let f be a real-valued function defined on the closed interval $[a, b]$. Show that if the set of Riemann sums for f is bounded, then f is bounded. By Riemann sum we mean a sum of the form

$$\sum_{i=1}^{n} f(t_i)(x_i - x_{i-1})$$

where $a = x_0 < x_1 < \cdots < x_n = b$ and $x_{i-1} \le t_i \le x_i$ for $1 \le i \le n$.

P1968-2. Given four points which are the vertices of a convex quadrilateral in the plane and five points inside the quadrilateral such that no three of the nine points are collinear, show that five of the nine points are the vertices of a convex pentagon.

P1968-3. Let $f : \mathbf{R}^n \mapsto \mathbf{R}^n$ be a differentiable function such that $f(tx) = tf(x)$ for $x \in \mathbf{R}^n$ and $t > 0$. Show that f is linear.

P1968-4. Find all integral solutions of the equation $2^x - 3^y = 1$ or of the equation $3^x - 2^y = 1$.

P1968-5. Let $\{f_n\}$ be a sequence of real-valued functions defined on **R**. Suppose that for each n, $\{x | f_n(x) \ne 0\}$ is bounded and that the sequence converges uniformly on **R** to the zero function. Show that

$$\lim_{n\to\infty} \int_{-\infty}^{\infty} f_n = 0.$$

P1968-6. Find two decreasing sequences $\{a_n\}$ and $\{b_n\}$ of positive numbers such that

$$\sum_{n=1}^{\infty} a_n = \infty \text{ and } \sum_{n=1}^{\infty} b_n = \infty, \text{ but } \sum_{n=1}^{\infty} c_n < \infty$$

where $c_n = \min\{a_n, b_n\}$.

P1968-7. Let $z_1, z_2, \ldots, z_n$ be complex numbers such that

$$|z_1| + |z_2| + \cdots + |z_n| = 1.$$

Show that for some $i_1, i_2, \ldots, i_k$, we have

$$|z_{i_1} + z_{i_2} + \cdots + z_{i_k}| \geq \frac{1}{\pi}.$$

P1968-8. Let n and k be positive integers. Suppose line segments are drawn joining each pair of n points and that each segment is painted blue or green. Are there k points such that all the line segments with end points among these k points are of the same color? Show that the answer is yes if n is large enough. Can you guess how large n must be?

Exam #4–1969

This competition was held at Rose-Hulman Institute of Technology, located in Terre Haute. The winning team from Valparaiso University consisted of Gerald Anderson, Charles Frank, and Charles Spear.

P1969-1. Prove that $\sin x \geq x - (x^2/\pi)$ if $0 \leq x \leq \pi$.

P1969-2. Suppose p, q, and r are positive integers no two of which have a common factor larger than 1. Suppose P, Q, and R are positive integers such that $\frac{P}{p} + \frac{Q}{q} + \frac{R}{r}$ is an integer. Prove that each of $\frac{P}{p}, \frac{Q}{q}$, and $\frac{R}{r}$ is an integer.

P1969-3. Determine whether $\sqrt[3]{25 + 5\sqrt{20}} + \sqrt[3]{25 - 5\sqrt{20}}$ is rational or irrational.

P1969-4. A ball is thrown into the air. The only forces acting are gravity (constant) and air resistance (proportional to the velocity). Which takes it longer, to go up, or to come down?

P1969-5. An equilateral triangle is circumscribed about an arbitrary triangle as shown in the figure below. Show without using calculus that the maximum area it can have is

$$\frac{1}{\sqrt{3}}\left(b^2 + c^2 - 2bc\cos\left(A + \frac{\pi}{3}\right)\right).$$

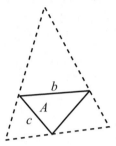

P1969-6. Assume that f has a continuous second derivative, that $a < b$, that $f(a) = f(b) = 0$, and that $|f''(x)| \le M$ on $a \le x \le b$. Prove that

$$\left|\int_a^b f(x)dx\right| \le \frac{M}{12}(b-a)^3.$$

Exam #5–1970

Held at Butler University, in Indianapolis, for the first of many occasions, this competition was again won by a team from Valparaiso University. The team consisted of Gerald Anderson, Jonathon Berke, and George Lueker. A team from Rose-Hulman came in second.

P1970-1. Evaluate

$$\lim_{n\to\infty}\left(\frac{1}{n+1} + \frac{1}{n+2} + \cdots + \frac{1}{2n}\right).$$

P1970-2. Let $f(x) = x^n + a_{n-1}x^{n-1} + \cdots + a_1x + a_0$ be a polynomial with real coefficients. Prove that any root c satisfies

$$|c| \le 1 + |a_{n-1}| + |a_{n-2}| + \cdots + |a_0|.$$

(Hint: consider $|c| \le 1$ and $|c| \ge 1$ separately.)

P1970-3. Squares $ABEF$, $BCGH$, $CDJK$, and $DALM$ are placed externally on the sides of a parallelogram $ABCD$, with X, Y, Z, and W the respective centers of those squares. Prove that the length from X to Z is the same as the length from Y to W, that the line segment from X

to Z is perpendicular to the line segment from Y to W, and finally that $XYZW$ is a square.

P1970-4. On the border of a disk select an even number, e, of points. Draw $e/2$ non-overlapping curves in the disk whose ends are the e dots. For instance, in the case $e = 10$ we may have the figure below. The curves cut the disk into $e/2 + 1$ regions. Prove that the regions can be colored with two colors in a way such that adjacent regions are colored differently.

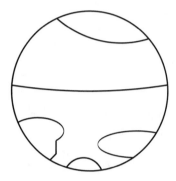

P1970-5. Find the smallest natural number n that has the following properties:

(a) Its base ten representation has a 6 as the last digit.

(b) If the last digit, 6, is erased and placed as the first digit in front of the remaining digits, then the resulting number is four times as large as the original number n.

Exam #6–1971

This competition was held at Earlham College, in Richmond. First place was won by a team from Rose-Hulman Institute of Technology—the first of their many wins—consisting of Tom Dehne, Bill Lipp, and Robert Klim. There was a tie for second place between teams from Earlham College and Goshen College.

There was a glossary of terms at the beginning of this exam, but this has been lost over time.

P1971-1. Let S be a set and let P be an equivalence relation on S. That is, for all x, y, z in S, xPx; $xPy \Rightarrow yPx$; and $xPy \wedge yPz \Rightarrow xPz$. For each subset A of S, define $\overline{A} = \{x \in S : \exists\, y \epsilon\, A \text{ such that } yPx\}$.

(a) Prove that $A \subset S$ and $B \subset S$ implies

    i. $A \subset \overline{A}$,

    ii. $\overline{\overline{A}} = \overline{A}$, and

    iii. $\overline{\overline{A} \cup \overline{B}} = \overline{A} \cup \overline{B}$.

(b) Let S be the set of points in the Cartesian plane. Define

$$R : (x_1, y_1) R(x_2, y_2) \text{ iff } y_1 - y_2 = 3(x_1 - x_2).$$

Granted that R is an equivalence relation on S, describe or sketch $\overline{A}$, where A is the unit circle with center at $(0,0)$. That is $A = \{(x,y) \in S : x^2 + y^2 = 1\}$.

P1971-2. Granted that the last three digits in the decimal representation of 7^{400} are 0, 0, 1, so that $7^{400} = 10^3 k + 1$ for some positive integer k, find the last three digits in the decimal representation of 7^{9999}.

P1971-3. Compute the derivative of $\frac{1}{1-e^{-1/x}}$. Then compute

$$\int_{-1}^{1} \left| \frac{e^{-1/x}}{x^2(1 - e^{-1/x})^2} \right| dx.$$

P1971-4. Let A be a countable subset of a Euclidean plane. Show that there exists a line in the plane which contains no point of A.

P1971-5. For any real number x, let $\lfloor x \rfloor$ denote the greatest integer in x. Thus, if n is the (unique) integer such that $n \le x < n+1$, then $\lfloor x \rfloor = n$. Let $x - \lfloor x \rfloor$ be called the fractional part of x.

(a) What is the limit of the fractional part of $(2 + \sqrt{2})^k$ as $k \to \infty$?

(b) What is the behavior of the fractional part of $(1 + \sqrt{2})^k$ as $k \to \infty$?

P1971-6. A set of eleven bank robbers decided to store their loot from a successful job in a safe (where else?) until the heat was off. Being somewhat but not completely trustful of one another, they decided to be able to open the safe when and only when a majority of the robbers was present. Therefore, the safe was provided with a number of different locks (the key to any one not being able to open any other), and each robber was given keys to some of the locks. How many locks were required, and how many keys had to be given to any one bank robber?

P1971-7. Is the following a Boolean algebra? Explain.

Given a non-empty set S together with a binary operation $\cap$ and a unary operation $*$. Suppose that

(a) for all A, B in S, $A \cap B = B \cap A$; and

(b) for all A, B, C in S, $(A \cap B) \cap C = A \cap (B \cap C)$; and

(c) for all A, B, C in S, $A \cap *B = C \cap *C \Leftrightarrow A \cap B = A$.

Exam #7–1972

This year's competition was held at DePauw University, located in Green-castle, just west of Indianapolis. The winning team was from Purdue University-Calumet, and it consisted of Reinhard Fritz, David Hasza, and Lawrence Kus.

P1972-1. Suppose that A, B, C, and D are square matrices. A and B are not invertible. What conditions on A and B are necessary for the existence of matrices X and Y such that $AX + BY = C$ and $BX + AY = D$?

P1972-2. If A is a square matrix such that $A^3 + 4A^2 + 3A + 2I = 0$, show that A is invertible (I denotes the identity matrix and 0 the matrix all of whose entries are zero.)

P1972-3. Which is bigger: e^{π} or π^{e}?

P1972-4. If $\sum_{n=1}^{\infty} a_n$ converges, where $a_n > 0$ and $a_n \neq 1$ for all n, does

$$\sum_{n=1}^{\infty} \frac{a_n}{1 - a_n}$$

converge? (Prove or give a counterexample.)

P1972-5. A card-shuffling machine always rearranges cards in the same way relative to the order in which they were given to it. All of the hearts arranged in order from ace to king were put into the machine, and then the shuffled cards were put into the machine again to be shuffled again. If the cards emerged in the order 10, 9, Q, 8, K, 3, 4, A, 5, J, 6, 2, 7, what order were the cards in after the first shuffle?

P1972-6. Find the points of intersection of the curves whose equations in polar form are

$$r = \cos(\theta/2) \quad \text{and} \quad r = \sin(\theta/2).$$

P1972-7. Can a group be a union of two proper subgroups?

Exam #8–1973

This competition was held at St. Mary's-of-the-Woods College, in Terre Haute. The team consisting of Tom Seilke, Tom Stocks, and John VanDrie from Wabash College won, with teams from Earlham College and Rose-Hulman taking second and third places respectively.

P1973-1. Evaluate $\lim\limits_{x \to \infty} x(e^{1/x} - 1)$.

P1973-2. A rectangle is inscribed in a sector of a circle of radius 1 as shown in the figure below. The central angle of the sector is a given angle θ, with $0 < \theta \le \pi/2$. Show that the maximum possible area for the rectangle is

$$\frac{1 - \cos\theta}{2\sin\theta}.$$

P1973-3. Let n be a fixed positive integer greater than 1. Determine the maximum value of $\prod_{i=1}^{k} n_i$, where n_i are positive integers whose sum is n. Note that $\prod$ is the product symbol and k is not fixed but can assume any value from 1 to n.

P1973-4. Evaluate

$$\lim_{n \to \infty} \int_0^1 \frac{ny^{n-1}}{1 + y} \, dy.$$

P1973-5. Let A and B be ideals of a commutative ring R. The quotient, $A : B$, of A by B is defined as $A : B = \{r \in R : rb \in A \text{ for all } b \in B\}$. Is $A : B$ an ideal of R? Prove or disprove.

P1973-6. If n is a positive integer, show that $a^{n+1} - n(a - 1) - a$ is divisible by $(a - 1)^2$.

P1973-7. Let A and B be square matrices. Prove that if AB is non-singular, then both A and B are non-singular.

P1973-8. Find all solutions in integers of $x^2 + y^2 = z^2$ with x, y, and z in arithmetic progression.

P1973-9. A group of 5 men contains 3 Democrats and 2 Republicans. Three men are chosen at random from the group. What is the probability that both of the Republicans were selected?

Exam #9–1974

This competition was held at Butler University for the second time, where the team of Stanley Lyness, Tom Sielke, and David Wilde from Wabash College won the competition.

P1974-1. Let

$$P(x) = (1+x)^{1000} + x(1+x)^{999} + x^2(1+x)^{998} + \cdots + x^{1000}.$$

(a) Find the coefficient of x^{50} in $P(x)$.

(b) What is the sum of all the coefficients of this polynomial?

P1974-2. Let S be a set with an associative multiplication, $(x, y) \rightarrow xy$. Suppose that for all x, y in S we have $x^3 = x$ and $x^2 y = yx^2$. Show that the multiplication is commutative.

P1974-3. For each positive integer m, find two distinct pairs of positive integers (n_1, N_1) and (n_2, N_2)(depending upon m) such that

$$(m^2 + 1)(n_i^2 + 1) = N_i^2 + 1 \qquad (i = 1, 2).$$

P1974-4. Determine whether at $(0,0)$ f achieves a local maximum, minimum, or neither, for the function $f(x, y) = (y^2 - x)(2y^2 - x)$.

P1974-5. Suppose that y is a continuously differentiable function of x which satisfies the condition $y(0) = 1$ and the inequality $\frac{dy}{dx} + e^x y + 1 \leq 0$. Show that y has a zero in the interval $[0, 3/4]$.

P1974-6. Examine the validity of the following conjecture: The series of positive terms $\sum_{n=0}^{\infty} a_n$ diverges if and only if the series $\sum_{n=0}^{\infty} a_n^2$ diverges.

Exam #10–1975

A team from Rose-Hulman won this competition, which was held at Wabash College. The winning team consisted of Robert E. Copus, Michael J. Dominik, and Barry W. Carlin.

P1975-1. Show that

$$-\frac{\pi}{2} < \sum_{n=1}^{\infty} \frac{a}{a^2 + n^2} < \frac{\pi}{2}.$$

P1975-2. A polygon having all its angles equal and an odd number of vertices is inscribed in a circle. Prove that it must be regular.

P1975-3. Given m lines in the plane, with no two parallel and no three concurrent; into how many components do they divide the plane? Prove your assertion.

P1975-4. The expression $a|b$ means that a divides b. Suppose that $2^m | (3^m - 1)$.

(a) Show that if $m \neq 1$, then m is even, and

(b) Show that $m = 1$, 2, or 4.

P1975-5. Suppose that $a > 0$ and f is continuous for $0 \leq x \leq a$. Define

$$g(x) = \int_x^a \frac{f(t)}{t}\, dt$$

for $0 < x \leq a$. Show that

$$\int_0^a g(x)\, dx = \int_0^a f(x)\, dx.$$

Exam #11–1976

This competition was held at DePauw University. The winning team of Jay Ponder, Tom Sellke, and Matthew Wyneken was from Wabash College. Teams from Rose-Hulman and Franklin College came in second and third, respectively.

P1976-1. Determine all polynomials $p(x)$ such that

$$p(x^2 - 1) = |p(x)|^2 - 1 \quad \text{and} \quad p(2) = 2.$$

P1976-2. Let n be a positive integer such that $n + 1$ is divisible by 12. Prove that the sum of all of the divisors of n is divisible by 12.

P1976-3. Let

$$f(x) = a_1 \tan x + a_2 \tan\left(\frac{x}{2}\right) + a_3 \tan\left(\frac{x}{3}\right) + \cdots + a_n \tan\left(\frac{x}{n}\right),$$

where $a_1, a_2, \ldots, a_n$ are real numbers and where n is a positive integer. Given that $|f(x)| \le |\tan x|$ for $x \in \{-\pi/2, \pi/2\}$, prove that

$$\left| a_1 + \frac{a_2}{2} + \frac{a_3}{3} + \cdots + \frac{a_n}{n} \right| \le 1.$$

P1976-4. How many zeroes does the function $f(x) = 3^x - 1 - 2x^2$ have on the real line? Prove that your answer is correct. Hint: You may need to know that $1 < \ln 3 < 1.1$.

P1976-5. Let A, B, and C be three non-collinear points in a rectangular coordinate plane with coordinates $(a_1, a_2), (b_1, b_2)$, and (c_1, c_2) respectively. Prove, using algebraic methods, that it is always possible to solve for the coordinates of the center of the circle containing A, B, and C.

P1976-6. Suppose that f is a real-valued function of a real variable, and that $f(x + y) = f(x)f(y)$ for all x and y, $f(1) \ne 0$, and $\lim_{x \to 0} f(x)$ exists. Prove that $\lim_{x \to 0} f(x) = 1$.

Exam #12–1977

This competition was held at Rose-Hulman and was also won by a team from there. The team consisted of Rich Priem, Jan Slupesky, and Bob Strickland.

P1977-1. Let f be a continuous function that maps the closed unit interval $J = [0, 1]$ into itself. Show that if $f(f(x)) = x$ for all x in J, then either f is strictly increasing on J or f is strictly decreasing.

P1977-2. Prove the following identities:

(a) $\cos^n x = \dfrac{1}{2^n} \displaystyle\sum_{k=0}^{n} \binom{n}{k} \cos(n - 2k)x$

(b) $\dfrac{1}{\pi} \displaystyle\int_0^\pi \cos^{2n} x \, dx = \dfrac{(2n)!}{2^{2n}(n!)^2}.$

P1977-3. Let $A = (a_{ij})$ be an $n \times n$ matrix of real numbers such that $\sum_{j=1}^{n} a_{ij} = 1$ for each i. Show that the matrix $A - I$ is not invertible.

P1977-4. Find all of the isometries of the set $\mathbf{R}$ of real numbers, that is, functions f from $\mathbf{R}$ into $\mathbf{R}$ with the property that the distance between any pair of points is the same as the distance between their images under f. Hint: Consider first the special case in which $f(0) = 0$.

P1977-5. Use the fact that $\sum_{n=1}^{\infty} 1/n$ diverges to obtain a proof that there are infinitely many prime numbers $p_1, p_2, \ldots$. Hint: For each integer N there is an integer r such that

$$\frac{1}{p_1^{a_1} p_2^{a_2} \cdots p_N^{a_N}} + \frac{1}{p_1^{b_1} p_2^{b_2} \cdots p_N^{b_N}} + \cdots + \frac{1}{p_1^{c_1} p_2^{c_2} \cdots p_N^{c_N}}$$

$$\leq \left(1 + \frac{1}{p_1} + \frac{1}{p_1^2} + \cdots + \frac{1}{p_1^r}\right)\left(1 + \frac{1}{p_2} + \frac{1}{p_2^2} + \cdots + \frac{1}{p_2^r}\right)$$

$$\cdots \left(1 + \frac{1}{p_N} + \frac{1}{p_N^2} + \cdots + \frac{1}{p_N^r}\right)$$

Exam #13–1978

This competition, held at Earlham College, was again won by a team from Wabash College. The team consisted of Kevin Fosso, Jay Ponder, and Matthew Wyneken. Teams from Rose-Hulman and Manchester College came in second and third, respectively. The exam came with instructions:

For problems 1 and 4 you need only show your computations, whereas problems 2, 3, 5, and 6 require more justification.

P1978-1. The number 10 is a base for the positive integers because every positive integer can be written uniquely as

$$d_n 10^n + d_{n-1} 10^{n-1} + \cdots + d_1 10 + d_0,$$

where each d_i is one of the digits 0, 1, 2, 3, 4, 5, 6, 7, 8, or 9. The number -2 is a base for all integers using the digits 0 and 1. For example 1101 represents -3, since

$$1(-2)^3 + 1(-2)^2 + 0(-2) + 1 = -3.$$

Find the representation in base -2 for the decimal number -2374.

P1978-2. Prove the following generalization of the Theorem of Pythagoras: If $ABCD$ is a 3-dimensional tetrahedron such that each of the triangles ABC, ABD, and ACD has a right angle at A, then

$$(\text{area} ABC)^2 + (\text{area} ABD)^2 + (\text{area} ACD)^2 = (\text{area} BCD)^2.$$

P1978-3. Let k be a positive odd integer and let $S(n) = \sum_{j=1}^{n} j^k$. Show that $2S(n)$ is divisible by $n + 1$.

P1978-4. Let $a_1, a_2, a_3, \ldots, a_n$ be real numbers, not necessarily distinct, and let $f(x) = \sum_{k=1}^{n} |x - a_k|$. For which value(s) of x is $f(x)$ smallest? What is the minimum value of $f(x)$? You need not give a formal justification for your answer.

P1978-5. Suppose that A, B, C, and D are four distinct points in the plane. Find the point O in the plane so that the sum of the distances from O to each of the four points is smallest, i.e., so that

$$|OA| + |OB| + |OC| + |OD|$$

is smallest.

P1978-6. Design an experiment with a fair coin for which the probability of success is $1/3$. A fair coin is a coin for which the probability of tossing heads and the probability of tossing tails are each $1/2$. As an illustration, we present the following experiment, for which the probability of success is $1/4$: Toss a fair coin twice. The experiment is a success if heads is tossed both times.

Exam #14–1979

The team of Mike Hall, Tim Drabik, and Tony Mazzoni from Rose-Hulman won this competition, which was again held at Butler University.

P1979-1. A piece of rope weighs four ounces per foot. It is passed over a pulley, and on one end is suspended a weight, and on the other end a monkey. The whole system is in equilibrium. The weight of the monkey in pounds is equal to the age of the monkey's mother in years. The age of the monkey's mother added to the age of the monkey is four years. The monkey's mother is twice as old as the monkey was when the monkey's mother was half as old as the monkey will be when the monkey is three times as old as the monkey's mother was when the monkey's mother was three times as old as the monkey. Each of the weight of the rope and the weight at the end is half again as much as the difference in weight between the weight and the weight plus the weight of the monkey. How long is the rope? Assuming the pulley to be negligibly small, what fractional part of the rope is on the same side of the pulley as the monkey when the system is at rest?

P1979-2. Graph the relation $\sin x = \sin y$ in the x, y-plane.

P1979-3.   Snow starts to fall in the forenoon and falls at a constant rate all day. At noon a snow plow starts to clear a highway. The velocity of the snow plow is such that it removes a constant volume of snow per unit of time. The plow goes a mile during the first hour. It goes a half mile during the second hour. What time did it start to snow? Give the time to the nearest minute.

P1979-4.   A hole six inches long is drilled through a sphere. (The hole is six inches long after it is drilled, and the hole is through the center of the sphere.) What is the volume of the part of the sphere that remains?

P1979-5.   Note that $1 = 1^{979}$, $2 = (-1)^{97} + \sqrt{9}$, $3 = 1^9 + |7 - 9|$, $4 = 1 + \sqrt{9} + \lfloor 7/9 \rfloor$.

Continue this through consecutive numbers k as far as you can. The rules of the game are as follows:

(i) The only constants you may use are 1, 9, 7, 9, which must be used exactly once and in that order.

(ii) Algebraic grouping symbols, operations, and functions may be used as often as desired. Note that rule (i) prohibits use of π or e, as well as use of higher order roots.

P1979-6.   Find all points with integer x and y coordinates which are strictly inside the polygon $ABCDEFGH$ (described below) and also strictly outside the circle centered at $(-4, -1)$ with radius $3/2$. Here is a description of the polygon:

$B = (4, -21), D = (11, -1), E = (-2, -2), F = (-4, 14)$

L_i ($i = 1$ to 8) are lines as follows:

L_1 goes through B and $(-6, -2)$

L_2 goes through B and has slope $-9/4$

L_3 goes through D and has slope $2/15$

L_4 goes through D and E

L_5 goes through E and F

L_6 goes through F and is perpendicular to $x - 16y = 4$

L_7 has equation $x + 7y = 4$

L_8 has a y-intercept of 6 and angle of inclination $45°$

A is the intersection of L_1 and L_8

C is the intersection of L_2 and L_3

G is the intersection of L_6 and L_7

H is the intersection of L_7 and L_8.

Exam #15–1980

This year, for the first time, the competition was held at Valparaiso University, located in Valparaiso. The winning team from Rose-Hulman consisted of Michael Call, Randall Ekl, and Douglas Englehart. A team from Butler University came in second.

P1980-1. Let $\{a_n\}$ be a sequence of positive numbers. If there exists a sequence $\{b_n\}$ of positive numbers and a constant $\alpha > 0$ such that

$$b_n \frac{a_n}{a_{n+1}} - b_{n+1} \geq \alpha,$$

show that the series

$$\sum_{n=1}^{\infty} a_n$$

is convergent.

P1980-2. Consider a balance that is used to measure loads of integral weights. The balance has two scales, a load scale and a weight scale. On the weight scale one can place only certain measuring weights. On the load scale one can place the load to be measured and any desired subset of the measuring weights. Show that with four suitably chosen weights one can measure the weight of any load whose weight is an integer between 1 and 40 pounds.

P1980-3. Suppose that f is a function that satisfies

$$f''(x) + f'(x)g(x) - f(x) = 0$$

for some continuous function g. Prove that if f vanishes at a and at some $b > a$, then f is identically zero on $[a, b]$.

P1980-4. Show that $\det(I + xy^t) = 1 + x^t y$ for any vectors x and y in $\mathbf{R}^n$. Note that t denotes transpose.

P1980-5. Find all positive functions f that satisfy

$$\frac{f(x)}{f(y)} \leq 2^{(x-y)^2}$$

for all x, y.

P1980-6. Show (a) that the integral

$$\int_0^{\infty} \frac{\sin x}{x}\, dx$$

is convergent. Then show (b) that

$$\frac{d}{d\alpha}\left(\int_0^\infty \frac{\sin \alpha x}{x}\, dx\right)^2 = 0.$$

P1980-7. If

$$S_n = \sum_{k=1}^n \frac{(-1)^{k+1}}{k},$$

show that

$$S_{2n} = \sum_{k=1}^n \frac{1}{n+k}.$$

Exam #16–1981

This competition, again won by a team from Rose-Hulman, was held at Indiana University-Purdue University in Indianapolis (IUPUI). The team consisted of Michael Call, Thomas Douglas, and Tim Drabik.

P1981-1. Express 3/8 as a base 7 fraction of form $0.a_0 a_1 a_2 a_3 \cdots$.

P1981-2. Before steers are introduced to a pasture, there is a given amount of grass per acre, and the grass keeps growing at a constant rate. If 12 steers take 16 weeks to deplete the grass on 10 acres, and if 18 steers take 8 weeks to deplete the grass on 10 acres, how many steers does it take to deplete the grass on 40 acres in 6 weeks?

P1981-3. A ball of radius 1 is in a corner touching all three walls. Find the radius of the largest ball that can be fitted into the corner behind the given ball.

P1981-4. The winning team of the World Series must win four games out of seven. Assuming that teams are equally matched, find the probabilities that the Series lasts

(a) exactly four games,

(b) exactly five games,

(c) exactly six games, and

(d) exactly seven games.

P1981-5. A man is standing atop a tall building. At a point 50 feet above his eye atop a building 100 feet away, a rock is dropped. If the man watches the rock fall, at what point in its descent is his head moving the fastest?

Assume that the buildings are arbitrarily high, that air friction is neglected, and that the acceleration of gravity is 32 ft/sec/sec.

P1981-6.   A destroyer is hunting a submarine in dense fog. The fog lifts for a moment, disclosing the submarine on the surface three miles away, upon which the submarine immediately descends. The speed of the destroyer is twice that of the submarine, and it is known that the latter will depart at once at full speed on a straight course of unknown direction. The wily captain of the destroyer sails straight to the point 2/3 of the way to the spot where the submarine was sighted and then sets out on a spiral course that is bound to make him pass directly over the submarine. What is the equation of this spiral? HINT: Use polar coordinates with the origin at the point where the submarine was sighted.

Exam #17–1982

This was the first time that the competition was held at Ball State University, located in Muncie. The winning team of Jeffery Baldwin, Randall Ekl, and J. Anthony Kirk was from Rose-Hulman.

P1982-1.   Find a cubic equation whose roots are the reciprocals of the roots of the equation $x^3 + ax^2 + bx + c = 0$, with $c \neq 0$.

P1982-2.   A subset $\{a_1, a_2, \ldots, a_k\}$ of the set $\{1, 2, \ldots, n\}$ is said to be separated if $a_{i+1} - a_i \geq 2$ for $i = 1, 2, \ldots, n-1$. For example, $\{2, 5, 7\}$ is a separated subset of $\{1, 2, \ldots, 8\}$, but $\{3, 4, 8\}$ is not. Show that the number of separated subsets of $\{1, 2, \ldots, n\}$, each having k elements is $\binom{n-k+1}{k}$.

P1982-3.   Given finitely many points in the plane situated so that any three of them are the vertices of a triangle of area ≤ 1. Show that all the points can be enclosed in a rectangle of area ≤ 4.

P1982-4.   Is the function

$$f(x) = \begin{cases} e^{-1/x^2}, & x \neq 0 \\ 0, & x = 0 \end{cases}$$

differentiable at x = 0? Prove your answer. Hint: You may use the fact that

$$\lim_{y \to \infty} y^r e^{-y} = 0$$

for any r.

P1982-5. A real-valued function f of a real variable is said to satisfy a Hölder condition with exponent α if there is a constant c such that $|f(x) - f(y)| \leq c|x - y|^{\alpha}$ for all x, y. Wherever these functions are used, α is restricted to be ≤ 1. Can you explain why?

P1982-6. The probability that the square of a positive integer (in decimal notation) ends with the digit 1 is $2/10$ because out of every 10 numbers those and only those ending with the digits 1 or 9 have squares ending with 1. What is the probability that the cube of a positive integer chosen at random ends with the digits 11? Prove your answer.

P1982-7. Find the volume of a torus (doughnut) of inner radius b whose cross-section by a plane through the axis is a semicircle of radius a, with its straight boundary parallel to the axis and curved boundary away from the axis.

Exam #18–1983

Rose-Hulman won the competition again this year. Its winning team consisted of Baron Gemmer, J. Anthony Kirk, and Tom Moss. The competition was held at the Indiana University, Bloomington, campus for the first time. Teams from Butler University and Manchester College tied for second place.

P1983-1. Find $\lim_{n \to \infty} \sqrt[n]{n!}$.

P1983-2. Suppose you repeatedly toss a fair coin until you get two heads in a row. What is the probability that you stop on the 10th toss?

P1983-3. Consider an isosceles right triangle with legs of fixed length a. Inscribe a rectangle and a circle inside the triangle as indicated in the figure below. Find the dimensions of the rectangle (and the radius of the circle) which make the total area of the rectangle and circle a maximum.

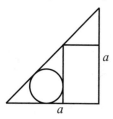

P1983-4. Show that every positive real number is a sum (possibly infinite) of a subset of the numbers $\{1, 1/2, 1/3, 1/4, \ldots\}$.

P1983-5. This problem is a repeat of problem #3 on EXAM #7.

P1983-6. Prove the "restricted" Fermat conjecture: For any integer $n > 2$, $x^n + y^n = z^n$ has no solutions in positive integers if $z < n$.

P1983-7. Let N be one more than the product of four consecutive positive integers. What can you say about N? Prove it.

P1983-8. Let $x_1, x_2, \ldots, x_n$ be (distinct) real numbers. Define polynomials $p_1(x)$, $p_2(x), \ldots, p_n(x)$ by

$$p_k(x) = \prod_{j \neq k} \frac{x - x_j}{x_k - x_j}, \qquad x \in \mathbf{R}.$$

Prove that $\sum_{k=1}^{n} p_k(x) = 1$ for all x in $\mathbf{R}$.

P1983-9. Suppose that $g : [0, 1] \to [0, 1]$ is a continuous function and that $g^m(x) = x$ (g composed with itself m times) for all x and for some positive integer m. Prove that $g^2(x) = x$ for all x.

Exam #19–1984

This competition was held at Rose-Hulman and a team from Rose-Hulman consisting of Todd Fine, Byron Bishop, and Dan Johnson won. Teams from Wabash College and Valparaiso University came in second and third, respectively.

P1984-1. What is the prime factorization of 1,005,010,010,005,001 ?

P1984-2. In the graph below, how many paths that never go up connect node A to node B? (Paths must follow edges indicated.)

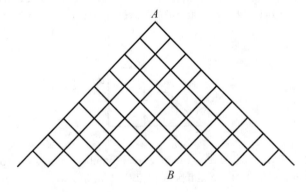

P1984-3. A very long hallway has 1000 doors numbered 1 to 1000; all the doors are initially closed. One by one, 1000 people go down the hall: the first person opens each door, the second person closes all doors with even numbers, the third person closes door 3, opens door 6, closes door 9, opens door 12, etc. That is, the nth person changes all doors whose numbers are divisible by n. After all 1000 people have gone down the hall: which doors are open and which are closed?

P1984-4. In Subsylvania, there is no paper money and there are only two kinds of coins, one worth 7 quanta, the other worth 12 quanta. Find the largest price for which it is not possible for a Subsylvanian customer to give exact change, or prove there is no such price. [Naturally, all prices are whole numbers of quanta. Note that it is possible for a Subsylvanian customer to buy an item costing 5 quanta (she gives the merchant a 12 quanta coin and receives a 7 quanta coin in exchange), but it is not possible for her to give the merchant 5 quanta in exact change.]

P1984-5. Choose two points at random in the interval $[0, 1]$. These two points cut the interval into three segments. What is the probability that these three segments can form the sides of a triangle? Note: "at random" means the two points are chosen independently and with uniform probability distribution. The probability of choosing a point in the interval $[a, b]$ is $b - a$.

P1984-6. Let f be a continuous function on $0 < x < \infty$ satisfying

$$f(1) = 5 \quad \text{and} \quad f\left(\frac{x}{x+1}\right) = f(x) + 2 \quad \text{for} 0 < x < \infty.$$

(a) Find $\lim_{x \to \infty} f(x)$.
(b) Prove that $\lim_{x \to 0^+} f(x) = +\infty$.
(c) Find all such functions f. (Part credit for finding some f.)

P1984-7. A train is being made up of Boxcars, Flat cars, and Hopper cars. Boxcars weigh 25 tons, and Flat cars and Hopper cars weigh 50 tons each. How many different trains of total weight 1000 tons (ignoring caboose and engines) can be made up? Trains are considered identical if and only if they have the same sorts of cars in the same order: e.g., HHB, HBH and FBH are distinct trains of total weight 125 tons.

28 Exams

Exam #20–1985

Held at Ball State University, this competition was again won by a team from Rose-Hulman. The team consisted of Todd Fine, Erick Friedman, and Dan Johnson. Teams from Rose-Hulman and Goshen College came in second and third, respectively. Notice that the contest rules for assigning second place were broken this year; it was the first of several in which this occurred.

P1985-1. Let α, β, γ, and δ be the roots of $x^4 + bx^3 + cx^2 + dx + e = 0$. Compute

$$(\alpha^2 + 1)(\beta^2 + 1)(\gamma^2 + 1)(\delta^2 + 1)$$

in terms of b, c, d, and e.

P1985-2. Prove: $3333^{4444} + 4444^{3333}$ is divisible by 7. Make up some more problems of this type.

P1985-3. Consider the set $U = \{1, 2, 4, 5, 8, 10, 11, 13, 16, 17, 19, 20\}$. In the figure below, each of the three paths—the solid, the dotted, and the dashed—represents a subgroup of order 6 of U under a certain binary operation. Three of the points of these paths have been labeled. Find all the others.

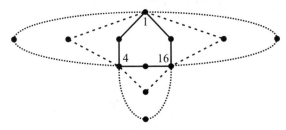

P1985-4. Each side of a square is subdivided into 101 segments. Find the number of different triangles that have their vertices at these 400 points of subdivision.

P1985-5. Let $a_1 = 0$ and $a_2 = 1$, and for $n \geq 3$,

$$a_n = (n - 1)(a_{n-1} + a_{n-2}).$$

Find

(a) a formula for a_n and

(b) $\lim_{n \to \infty} \frac{a_n}{n!}$.

P1985-6. Given $n + 1$ integers between 1 and $2n$ inclusive, prove that one of them must be a multiple of some other one.

P1985-7. Let a and b be positive constants with $b > 1$. Given that $x + y = 2a$ and all values of x between 0 and $2a$ are equally likely, find the probability that

$$xy > \frac{(b^2 - 1)a^2}{b^2}.$$

Exam #21–1986

This competition was held at Butler University. The winning team consisting of Erich Friedman, John Hoffman, and Dan Tretter was from Rose-Hulman. The second place team was also from Rose-Hulman, and the third place team was from Valparaiso University.

P1986-1. This is a repeat of problem #1 on EXAM #14.

P1986-2. Let $\{b_1, b_2, b_3\}$ be an orthonormal basis for $\mathbf{C}^3$ and let $\phi \in L(\mathbf{C}^3, \mathbf{C}^3)$ be given by

$$\phi(b_1) = 2b_2, \quad \phi(b_2) = 2ib_1, \quad \text{and} \quad \phi(b_3) = (1 + i)b_3.$$

Find the eigenvalues, if any, for ϕ.

P1986-3. Evaluate i^i.

P1986-4. N points are chosen on a circle so that when all segments are drawn between all pairs of points, no three segments intersect at the same point. In how many points do these segments intersect in the interior of the circle?

P1986-5. Find an equation with integral coefficients one of whose roots is $\sqrt{2} + \sqrt[3]{3}$.

P1986-6. Let $f : \mathbf{R}^2 \to \mathbf{R}$ be given by

$$f(x_1, x_2) = \begin{cases} \dfrac{x_1 x_2^2}{x_1^2 + x_2^4}, & \text{if } (x_1, x_2) \neq (0, 0); \\ 0, & \text{if } (x_1, x_2) = (0, 0). \end{cases}$$

(a) Is f continuous at $(0, 0)$? Justify your answer.

(b) Is f differentiable at $(0, 0)$? Justify your answer.

Exam #22–1987

This competition was held at Butler University. The team consisting of Daniel W. Johnson, Brenton Young, and Daniel Kniep from Rose-Hulman won the competition. A team from Indiana University, Bloomington, came in second, and a team from Wabash College came in third.

P1987-1. A positive integer n is called composite if there are positive integers $p \neq 1$ and $q \neq 1$ so that $n = pq$. Find a sequence of 10 consecutive positive integers each of which is composite and less than 1,000,000.

P1987-2. John's job at the Acme Cannonball Factory is to stack the cannonballs (which are 6-inch diameter spheres) neatly into tetrahedral piles. For example, using 4 cannonballs, John can make a tetrahedral pile with 2 cannonballs on each edge: three balls forming the bottom triangle and one in the center on the top. Find a formula for the number of cannonballs in a tetrahedral stack whose base is an equilateral triangle with n balls on each edge.

P1987-3. Experiments have determined that when a particular steel ball is bounced on a hard surface, it bounces to half its original height. For example, if it is dropped from a height of 6 feet, it will bounce to 3 feet. Assuming that the ball obeys this law exactly, for what length of time will the ball continue to bounce if it is dropped from a height of 16 feet (or will it bounce forever)? [Recall from calculus that since the acceleration due to gravity is 32 ft/sec/sec, an object failing to the ground from height h (in feet) or bouncing from the ground to height h requires $\sqrt{h}/4$ seconds to do so.]

P1987-4. Ten seniors who share a house decide to exchange graduation presents. They each put their name into a hat, mix the name cards thoroughly, and draw a card out at random. What is the probability that none of the ten draws his or her own name?

P1987-5. Euclidean four-space is R^4 with

$$|XY| = \left((y_1 - x_1)^2 + (y_2 - x_2)^2 + (y_3 + x_3)^2 + (y_4 + x_4)^2\right)^{1/2}.$$

(a) Find five points in the unit ball of Euclidean four-space that are as far from each other as you can make them. That is, find points A, B, C, D, and E such that

$$|OA| \leq 1, \quad |OB| \leq 1, \quad |OC| \leq 1, \quad |OD| \leq 1, \quad \text{and} \quad |OE| \leq 1,$$

where O is the origin, such that

$$\min\{|AB|, |AC|, |AD|, |AE|, |BC|, |BD|, |BE|, |CD|, |CE|, |DE|\}$$

is as large as you can make it.

(b) If your answer is best possible, prove that it is.

P1987-6. Jane begins a journey at Quito, Ecuador (at 80° west longitude on the equator) and flies steadily northwest.

(a) How far does she fly before reaching the North Pole?

(b) How many times does she cross the Greenwich Meridian (longitude 0°) in the journey?

(Note: You should suppose that the earth is a perfect sphere of radius 4000 miles, that there is no effect of wind, etc., that compass headings are exactly accurate, and that Jane's plane has an unlimited amount of fuel available.)

Exam #23-1988

This competition was held at Butler University. The contest was won by a team of two students from Indiana University, Bloomington: Radu Tudorica and Kevin Pilgrim. A team from IUPUI came in second, and a team from Purdue University came in third.

P1988-1. Going at top speed, Grand Prix driver x leads his rival y by a steady three miles. Only two miles from the finish, x runs out of fuel. Thereafter x's deceleration is proportional to the square of his remaining velocity, and, in the next mile, his speed exactly halves. Who wins and why?

P1988-2. Let $(G, *)$ be a group with the following cancellation rule: $x * a * y = b * a * c$ implies $x * y = b * c$ for all x, y, a, b, and c in G. Prove that G is Abelian, i.e., that * is commutative.

P1988-3. Define

$$T_k(x) = \prod_{i=1, i \neq k}^{n} (x - i),$$

and let $P(x)$ be a polynomial of minimum degree in which $P(k) = 5T_k(k)$, $k = 1, 2, \ldots, n$. If s and t are both integers such that $1 \leq s \leq n$ and $1 \leq t \leq n$, prove that $\int_s^t P(x)\, dx = 0$.

P1988-4. let S_n be the sum of the squares of the first n positive odd integers. What is the units digit of S_{12345}? Prove your answer.

P1988-5. Find the shaded area of the figure below, where the interior circular arcs have their centers on the outer circle. (This was a proposed problem in the then current issue of *The Journal of Recreational Mathematics*. Students were encouraged to send in solutions they liked.)

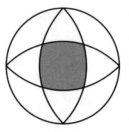

P1988-6. Show, with proof, how to construct with unmarked straightedge only, a perpendicular from the point P to the line L, as shown in the figure below. (This is problem 13 in the 1988 issue of *The Old Farmer's Almanac*. A solution to this plus several other problems could have won the solver 50 dollars.)

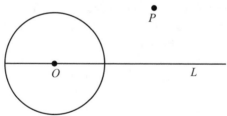

P1988-7. A fair coin is tossed ten times. Find the probability that two tails do not appear in succession.

Exam #24–1989

This competition was held on the Indiana University, Bloomington, campus. A team from Purdue University won the competition. The members of this team were Alex Gottlieb, Aaron Weindling, and Nung Kwan Yip. Teams from Earlham College and Wabash College came in second and third, respectively.

P1989-1. Three students A,B, and C compete in a series of tests. For coming first in a test, a student is awarded x points; for coming second,

y points; for coming third, z points. Here x, y, and z are positive integers with $x > y > z$. There were no ties in any of the tests. Altogether A accumulated 20 points, B 10 points, and C 9 points. Student A came in second in the algebra test. Who came in second in the geometry test?

P1989-2. Describe all sequences $\{x_1, x_2, \ldots, x_n\}$ of real numbers that satisfy

$$|x_m - x_n| \leq \frac{2mn}{m^2 + n^2}$$

for all positive integers m and n.

P1989-3. Show that the area of the shaded triangle formed by the side-trisectors of a triangle (as shown) is $1/7$ of the area of the whole triangle.

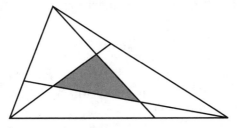

P1989-4. Two coins are given. One is fair [Prob(heads) = 1/2] and the other is biased with Prob(heads) = 2/3. One of the coins is tossed once, resulting in heads. The other is tossed three times, resulting in two heads. Which coin is more likely to be the biased one?

P1989-5. Two identical pipes have elliptical cross-sections with semi-axes a and b. The pipes intersect at right angles as shown below. Find the volume of their intersection.

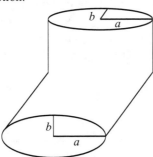

P1989-6. Find the determinant of the matrix

$$\begin{pmatrix} 1 + x_1y_1 & x_1y_2 & \cdots & x_1y_n \\ x_2y_1 & 1 + x_2y_2 & \cdots & x_2y_n \\ \vdots & \vdots & \ddots & \vdots \\ x_ny_1 & x_ny_2 & \cdots & 1 + x_ny_n \end{pmatrix}.$$

P1989-7. What are the dimensions of Smith's ranch, described in the following conversation?

Smith. Down in Todd County, which is a 19×19 miles square, I have a ranch. It is rectangular, not square, the short side and the long side both measuring a whole number of miles.

James. Hold on a minute! I happen to know the area of your ranch. Let me see if I can figure out its dimensions. (He figures furiously.) I need more information. Is the short side strictly less than half the long side? (Smith answers his question.)

James. Now I know the dimensions.

Kathy. I, too, know the area of your ranch, and although I did not hear your answer to James's question, I, too, can tell you the dimensions.

Bill. I did not know the area of your ranch, but having heard this entire conversation, I can now figure it out.

Exam #25–1990

Beginning in 1990, significantly more information was collected about the competitions. This year, the contest was held at Purdue University (located in West Lafayette) for the first time, on March 31, 1990 in connection with the spring meeting of the Indiana Section of the MAA. Twenty-four teams from twelve colleges participated. The winning team from Rose-Hulman consisted of Joel Atkins, John O'Bryan, and Kevin O'Bryant. They scored 51 out of 60 points on the following examination. The second place team from Purdue University consisted of Mark Sepanski, Peter Sepanski, and Nung Kwan Yip. The third place team from Indiana University consisted of Urmi Bhatacharya, Lucia Demetrios, and Radu Tudorica.

P1990-1. Find all positive integers which are one more than the sum of the squares of their base ten digits. For example, $35 = 1 + 3^2 + 5^2$.

P1990-2. A digraph in a word is an ordered pair of consecutive letters; a word with n letters has $n - 1$ digraphs. How many ways can the letters

I, N, D, I, A, N, A be arranged so that no digraph is repeated? (Thus, the arrangement A, N, D, I, I, N, A is counted, but not I, A, N, N, I, A, D, which contains "I, A" twice.)

P1990-3. Find all real functions f such that, for all real x,

$$f(x+2) = f(x) \quad \text{and} \quad f'(x) = f(x+1) - 2.$$

P1990-4. Let $ABCDEFG$ denote a regular heptagon with side 1. By connecting the vertices $ACEGBDFA$, in that order, we create a new regular heptagon $HIJKLM$ with side r. (See the figure below). Determine r.

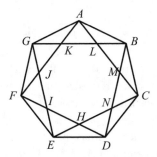

P1990-5. Evaluate the following limit.

$$\lim_{n \to \infty} n^{-2} \sum_{i=1}^{n} \sum_{j=1}^{n^2} \frac{1}{\sqrt{n^2 + ni + j}}$$

P1990-6. A regular hexagon of side 1 is inscribed in the intersection of two identical parabolas, oriented (in opposite directions) with their axes parallel to the y-axis (see the figure below). Find the area of the (shaded) region inside the parabolas and outside the hexagon.

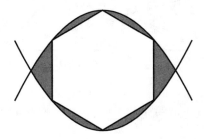

Exam #26–1991

The contest was held at Anderson University, located in Anderson, on March 23, in connection with the spring meeting of the Indiana Section of the MAA. Twenty-three teams from fourteen colleges participated. The winning team from Rose-Hulman consisted of Jonathan Atkins, John O'Bryan and Mark Roseberry. The second place team from Purdue University consisted of Alex Gottlieb, Peter Sepanski and Boon-Lock Yeo. The third place team from Goshen College consisted of Dave Cooper, Kathy Steiner and Ken Wenger.

P1991-1. Find the area of that portion of the xy-plane which is enclosed by the curve with equation

$$|2x - 1| + |2x + 1| + \frac{4|y|}{\sqrt{3}} = 4.$$

P1991-2. Between 1 and 1,000,000 inclusive, which are more numerous, those integers whose base-ten representations contain a "1" or those that do not? How many of each type are there?

P1991-3. P, Q and R are arbitrary points on sides BC, DA and CD respectively, of the parallelogram $ABCD$, illustrated below. A is joined to P, P to Q, Q to B, B to R and R to A to form a star-pentagon $APQBR$. The regions inside the parallelogram and outside the star-pentagon are colored red (r). The pentagonal region bounded by the sides of the star-pentagon is colored blue (b). Show that the red area minus the blue area is independent of the choice of the points P, Q and R.

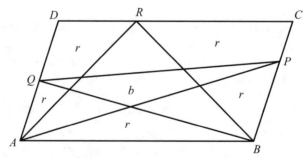

P1991-4. If a, b and c are integers and if $a+b+c$ is even, find an integer n (in terms of a, b and c) such that $ab + n$, $bc + n$ and $ca + n$ are all squares of integers.

P1991-5. (a) Show that there exist infinitely many twosomes of consecutive positive integers each of which can be written as the sum of two nonzero perfect squares. For example, $17 = 1^2 + 4^2$, $18 = 3^2 + 3^2$ and $25 = 3^2 + 4^2$, $26 = 1^2 + 5^2$.

(b) Find three consecutive positive integers each of which can be written as the sum of two nonzero perfect squares. Are there infinitely many such trios?

(c) Do there exist four consecutive positive integers each of which can be written as the sum of two nonzero perfect squares?

P1991-6. The harmonic mean of a set of positive numbers is the reciprocal of the arithmetic mean (ordinary average) of the reciprocals of the numbers. Find $\lim_{n \to \infty}(H_n/n)$, where H_n is the harmonic mean of the n positive integers $n + 1, n + 2, n + 3, \ldots, n + n$.

P1991-7. The centroid of a set of n points $\{(x_i, y_i), i = 1, 2, \cdots, n\}$, in the xy-plane is given by $\left(\frac{1}{n}\sum_1^n x_i, \frac{1}{n}\sum_1^n y_i\right)$. A lattice point in the xy-plane is a point whose coordinates are integers. Given nine lattice points in the xy-plane, show that some three of the points have a lattice point as centroid.

Exam #27–1992

The contest was held at The University of Indianapolis on April 11, in connection with the spring meeting of the Indiana Section of the MAA. There were 24 teams with a total of 67 participants from 13 colleges. The winning team from Rose-Hulman consisted of Jonathan Atkins, Tony Hinrichs, and Kevin O'Bryant. The second place team from Indiana-Purdue University at Fort Wayne consisted of Jeff Gerardot, Aaron Pesetski, and Mark Pragan. The third place team from Purdue University consisted of Peter Sepanski and Pok-Yin Yu.

P1992-1. A car rode over an ant on the pavement. The ant stuck to the tire for one revolution and then was deposited back onto the pavement. Assuming that the radius of the tire is one foot, find the length of the curve traveled by the ant between its death and its final resting place.

P1992-2. At a movie theater, n patrons have lined up to buy tickets. The ticket seller calls a patron viewable if he (she) is taller than all the people in front of him (her) in line; otherwise he (she) is hidden. Given that no two patrons are precisely the same height, find the average number of viewable patrons among all possible permutations of the patrons.

P1992-3. A collection of n gossips each knows a unique tidbit of scandal not known to any of the others. They communicate by mailing letters. Of course each gossip will share all of the scandal he (she) knows at that time whenever he (she) sends a letter. Find, with proof, the minimum number of letters that can suffice to share all of the scandal.

P1992-4. Identify all finite groups G of order n having at least 2^{n-2} proper subgroups.

P1992-5. Evaluate $\int_{-\infty}^{\infty} e^{-x^2}\, dx$.

P1992-6. For $A = \begin{pmatrix} 19 & 5 \\ -30 & -6 \end{pmatrix}$

(a) find the eigenvalues of A.

(b) Find four different integral matrices B so that $B^2 = A$. (Hint: How do the eigenvalues of B relate to the eigenvalues of A?)

(c) For an arbitrary 2×2 matrix A, what is the maximum number of integral matrices B that can satisfy $B^2 = A$?

P1992-7. Evaluate

$$\lim_{x \to 0^+} \frac{x}{\sqrt{1 - e^{-2x^2}}}.$$

P1992-8. We all know that the lengths 3, 4, and 5 form the sides of a right triangle. Notice that 3 and 4 are consecutive integers. Determine all other integral right triangles whose legs are consecutive integers.(The hypotenuse must also be an integer, but need not be consecutive as 5 happens to be.)

Exam #28–1993

The contest was held at Saint Mary's College, located at the northern edge of South Bend, on April 24, in connection with the joint spring meeting of the Indiana Section of the MAA with the Illinois and Michigan Sections. There were 30 teams, with representation from all three states. The winning team from Rose-Hulman consisted of Jon Atkins, Tony Hinrichs, and Nick Tallyn. The second place from from Indiana-Purdue University, Fort Wayne, consisted of Joel Holcombe, Brian Johnson, and Aaron Pesetski. The third place team from Calvin College, Michigan, consisted of Alan Baljeu, Mflce Bolt, and Tim Hoflebeek.

P1993-1. Let $P(x)$ be a real cubic polynomial for which $P'(x)$ has distinct real zeros. Prove that

$$\frac{P'''(x)}{P'(x)} < \frac{1}{2}\left(\frac{P''(x)}{P'(x)}\right)^2$$

for all x for which $P'(x) \neq 0$.

P1993-2. Let f and g be mappings from the set A to itself for which $f(g(f(a))) = g(a)$ and $g(f(f(a))) = f(a)$ for all a in A. Prove that $f = g$.

P1993-3. Let A be a square matrix of rank 1 and trace 1. Prove that $A^2 = A$.

P1993-4. For $n > 1$, a permutation $a_1, a_2, \ldots, a_n$ of $\{1, 2, \ldots, n\}$ is "orderly" if, for each $i = 1, 2, \ldots, (n-1)$, there is a $j > i$ for which $|a_j - a_i| = 1$. (For example, for $n = 4$, the permutation $1, 4, 2, 3$ is orderly whereas $3, 1, 4, 2$ is not.) How many permutations of $\{1, 2, \ldots, n\}$ are orderly? HINT: What are the possible values for a_1?)

P1993-5. Let $a_1 = 1$ and $a_{i+1} = \sqrt{a_1 + a_2 + \cdots + a_i}$, for $i > 0$. Determine $\lim\limits_{n \to \infty}\left(\frac{a_n}{n}\right)$.

P1993-6. Prove that 2^{n+1} is a factor of $\lceil(\sqrt{3}+1)^{2n}\rceil$ for all positive integers n. (Here $\lceil x \rceil$ denotes the smallest integer not less than x.)

P1993-7. Find all integers A, B, C, D, E ($A \leq B \leq C \leq D \leq E$) which, when added in pairs, yield only the sums 401, 546, 691, and 836.

P1993-8. In the figure below, $ABCD$ is a rectangle. Find the area of the parallelogram $abcd$.

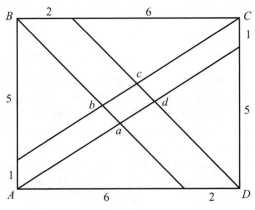

Exam #29–1994

The contest was held at Depauw University, on March 19, in conjunction with the spring meeting of the Indiana Section of the MAA. Twenty teams from eleven colleges participated. The winning team from Wabash College consisted of Faisal Ahmed, Avijit Sarkar, and Faisal Syud. The second place team from Rose-Hulman consisted of Jonathan Atkins, Nick Tallyn, and James Moore. The third place team from Purdue University consisted of Ilya Gluhovsky, Brian Singer, and William Stier. (These problems were used with the permission of *Crux Mathematicorum*, © 1975, 1976, 1977.)

P1994-1.   Calculate
$$\lim_{x \to \infty} \int_0^x e^{t^2 - x^2} \, dt.$$

P1994-2.   Show that for positive integers n, $\sqrt[n]{n} < 1 + \sqrt{2/n}$.

P1994-3.   If $x > y > 0$, show that
$$\ln x^2 - \ln y^2 < \frac{x^2 - y^2}{xy}.$$

P1994-4.   The sides of a triangle have lengths 4, 5, and 6. Show that one of its angles is twice another.

P1994-5.   Find all sets of five positive integers whose sum equals their product.

P1994-6.   Show that if one of the coordinates of the center of a circle is irrational, then the circle contains at most two points both of whose coordinates are rational.

Exam #30–1995

The contest was held at Tri-State University, located in the far northeastern community of Angola on March 31, in conjunction with the spring meeting of the Indiana Section of the MAA. Twenty-three teams from eleven colleges participated. The winning team from IU, Bloomington, consisted of Daniel Bliss, Matt Laue, and Seth Patinkin. The second place team from Wabash College consisted of Eham Ahmed Chowdhury, Matthew Gladden, and Shirish Ranjit. The third place team from Rose-Hulman consisted of Jarnie Kawabata, Rick Mohr, and Nick Tallyn.

P1995-1. Let z_1 and z_2 be complex numbers. Prove that

$$\left|z_1 + \sqrt{z_1^2 - z_2^2}\right| + \left|z_1 - \sqrt{z_1^2 - z_2^2}\right| = \left|z_1 + z_2\right| + \left|z_1 - z_2\right|.$$

P1995-2. Find a closed formula for the sum

$$\sum_{k=0}^{n} \binom{3n}{3k}.$$

P1995-3. Let p be an odd prime. Prove that the integer part of

$$(\sqrt{5} + 2)^p - 2^{p+1}$$

is divisible by $20p$.

P1995-4. Prove that

$$4(x^2 + x + 1)^3 - 27x^2(x + 1)^2 = (x - 1)^2(2x + 1)^2(x + 2)^2.$$

P1995-5. Let n be any integer greater than 1. Prove that

$$1^1 \cdot 2^2 \cdot 3^3 \cdots n^n < \left(\frac{2n + 1}{3}\right)^{\frac{n(n+1)}{2}}.$$

P1995-6. ABC is an equilateral triangle with each side of length a. L is a line parallel to BC and b units below BC. A solid S is generated by rotating the triangle ABC about the line L; find the volume of S.

P1995-7. The vertices of a triangle ABC have coordinates $(a \cos \alpha_i, a \sin \alpha_i)$, $i = 1, 2, 3$. Find the coordinates of the orthocenter of triangle ABC.

P1995-8. Find the coefficient of x^{2n} in the expansion of $\frac{(1+x)^n}{(1-x)^3}$ in closed form.

Exam #31–1996

The contest was held at Butler University on March 29, in conjunction with the spring meeting of the Indiana Section of the MAA. Twenty-five teams from eleven colleges participated. The winning team from Rose-Hulman consisted of Rick Mohr, Nick Tallyn, and James Moore. Indiana University, Bloomington, Purdue University, and Wabash College tied for second place. The IU team consisted of Seth Padnkin and Richard Swartz. The Purdue University team consisted of Dan Crosby, Lukito Muliadi, and Tze Chao, Ng. The Wabash College team consisted of Alexander Radnovich, Andrew Jones, and David Whittaker.

P1996-1. The San Francisco 49ers derive their name from the last two digits of the year of the California Gold Rush, 1849. Note that both numbers are perfect squares ($49 = 7^2$ and $1849 = 43^2$). How many other four-digit squares have this property, that their last two digits also form squares (count 00, 01, 04, and 09 as squares)?

P1996-2. Find the largest possible area of a pentagon with five sides of length 1 and a right interior angle.

P1996-3. A hundred armed rats enter a bar containing a hundred fat cats. After the cats are lined up against the wall, the first rat marches down the line taking a dollar from every cat. On his return, he passes the $100 that he collected to the second rat who proceeds to give a dollar to every second cat. On his return, he passes the remaining $50 to the third rat who proceeds to take a dollar from every third cat. This continues with the rats alternately giving and taking until the 100th rat gives a dollar to the 100th cat. At this point the rats and the money they have taken disappear in a cloud of smoke.

(a) How many cats profited? lost money? broke even from this?

(b) Which cat(s) profited the most and how much did they get?

(c) How much money did the rats leave with?

P1996-4. A square with sides parallel to the coordinate axes is inscribed in the region
$$\{(x, y) : x, y > 0, \ y \le 3x - x^3\}.$$

If its area is written as $\sqrt[3]{A} + \sqrt[3]{B}$, where A and B are integers, find $A + B$.

P1996-5. In triangle ABC, $\angle A = 90^\circ$ and $AB = AC = 2$. If curve γ joins points of AB and AC to bisect the area of triangle ABC, find the length of the shortest possible such curve.

P1996-6. Two couples each agree to have exactly n children. They plan to eventually pair off all their children in marriage so that sons and daughters of the first couple marry daughters and sons of the second couple. Assuming that sons and daughters are equally likely to be born, find the a priori probability p that such an arrangement is mathematically possible. Using Wallis' product, Sterling's formula, or any other well-known result, determine $\lim_{n\to\infty} (p_n)(\sqrt{n})$.

P1996-7.   Call two datasets "statistically equivalent" if they have the same number of elements, the same mean, and the same standard deviation. Find all datasets of positive integers (in any order) which are statistically equivalent to $\{1, 9, 9, 6\}$.

P1996-8.   Find all continuous functions $f(x)$ whose graph G (of $y = f(x)$) has the following property: For each chord C of G, if C's projection onto the x-axis has length d^2, then C's midpoint lies d units above G.

Exam #32–1997

The contest was held at Franklin College, on March 14, in conjunction with the spring meeting of the Indiana Section of the MAA. Twenty-five teams from thirteen universities participated. The winning team from Wabash College consisted of Jun Ma, Robert Dirks, and Abishai Daniel. Teams from Rose-Hulman and Purdue University placed second and third, respectively. The Rose-Hulman team consisted of Kyle Lacey, Chris Prince, and Tyson Patterson. The Purdue team consisted of Dan Crosby, Tze-Chao Ng, and Elad Harel.

P1997-1.   In Indiana, license plates consist of 2 digits followed by a letter and then 4 more digits. Find the probability of getting a license plate in which the last four digits are nondecreasing.

P1997-2.   It is well known that $1 + 2 + 3 + 4 + \cdots + n = n(n+1)/2$ for any positive integer n. Find a similar expression for the sum

$$1 + (1 + 2) + (1 + 2 + 3) + \cdots + (1 + 2 + 3 + \cdots + n).$$

P1997-3.   Find the smallest positive number A so that

$$\frac{21}{|z^4 - 5z^2 + 6|} \leq A$$

for every complex number z on the circle $|z| = 3$. Prove that A is the smallest such number.

P1997-4.   Find all solutions (x, y) of the equation $x^y = y^x$ for real numbers $x, y > 0$.

P1997-5.   Given the following set of axioms:

   (1) Every line contains exactly four points.

   (2) No pair of points may be on more than one line.

(3) Each point must be on at least four lines.

(4) Not all the points are on one line.

(5) There is at least one line.

(6) If a point is not on a given line, then the point must be on exactly one line that has no points in common with the given line

(a) What is the fewest number of points and lines necessary to satisfy these six axioms?

(b) If n were substituted for "four" in (1) and (3), what is the fewest number of points and lines necessary to satisfy these six axioms?

P1997-6. The usual way of measuring the length of a vector $x = (x_1, x_2, \ldots, x_n)$ in n-dimensional Euclidean space is the Euclidean norm

$$\|x\|_2 = \sqrt{\sum_{i=1}^{n} x_i^2}.$$

But there are other norms which can be used, such as

$$\|x\|_1 = \sum_{i=1}^{n} |x_i| \quad \text{and} \quad \|x\|_\infty = \max_{1 \le i \le n} |x_i|.$$

(a) We know that the set of points $\{x \epsilon R^2 : \|x\|_2 = 1\}$ is the unit circle. Find each of the following sets of points:

$$\{x \in R^2 : \|x\|_1 = 1\} \quad \text{and} \quad \{x \in R^2 : \|x\|_\infty = 1\}.$$

(b) The triangle inequality $\|x + y\| \le \|x\| + \|y\|$ holds for all norms. For the 2-norm, equality only holds if x and y are colinear. Determine when equality holds for the 1-norm and the ∞-norm.

P1997-7. A farmer has a trough 4 feet long with semicircular cross sections that are 2 feet in diameter. The top of the trough is parallel to the ground. Initially, the trough is full of water. If the farmer tilts the trough (along the side) at an angle α with the horizontal, express the volume (in terms of α) of the remaining water.

Exam #33–1998

This was the largest ICMC to date. Twenty-nine teams from 22 universities and schools participated in this ICMC, held at Ball State University. The winning team from Wabash College consisted of Robert Dirks,

Abishai Daniel, and Jun Ma. Teams from Rose-Hulman and the University of Evansville placed second and third, respectively. The Rose-Hulman team consisted of Kyle Lacey, Randy Motchan, Matt Lepinski, and the Evansville team consisted of Siddartha Naidu, Robert Linne, Hiten Sonpal.

P1998-1. Let C be a circle with diameter AB. Let P be any point of segment AB. Let C, D, E be distinct points of C all on the same side of line AB such that $DP \perp AB$ and $\angle CPD = \angle DPE$. Show that $(PD)^2 = PC \cdot PE$.

P1998-2. Show that an integer n with final (decimal) digit u is divisible by 7 if and only if $\frac{n-u}{10} - 2u$ is divisible by 7.

P1998-3. Prove that an odd integer $n > 1$ is composite iff there exist nonnegative integers p and q such that $n = p^2 - q^2$ with $p - q > 1$.

P1998-4. Let $f_0 = 0$, $f_1 = 1$, and $f_k = f_{k-1} + f_{k-2}$ be the Fibonacci numbers.

(a) Show that the points $\mathbf{F} = \{(f_k, f_{k+1})\}$ lie on two hyperbolas: $y^2 - xy - x^2 = \pm 1$.

(b) Show that the only points on the hyperbolas $y^2 - xy - x^2 = \pm 1$ with nonnegative integer coefficients are points in $\mathbf{F}$.

P1998-5. Let f_0, f_1, f_2 be three nonnegative increasing real-valued functions defined on the nonnegative real numbers with $f_i(0) = 0$ and $f_0(x) \le f_1(x) \le f_2(x)$. For any nonnegative x, let $R(x)$ denote the rectangle whose vertices are $(0,0), (x,0), (0, f_1(x))$, and $(x, f_1(x))$ Then, f_1 is said to bisect f_0 and f_2 in area if for every nonnegative x, the area of the portion of the $R(x)$ lying between the curves $y = f_0(x)$ and $y = f_1(x)$ has the same area as the portion of $R(x)$ lying between the curves $y = f_1(x)$ and $y = f_2(x)$. If $f_0(x) = x^\beta$ and $f_1(x) = \alpha x^\beta (\alpha > 1, \beta > 0)$, and f_1 bisects f_0 and f_2 in area, then what is $f_2(x)$?

P1998-6. Let $X_n = \{1, 2, \dots, n\}$. A permutation of n objects is a 1-1 function, φ, from X_n onto itself. φ is called a derangement if $\varphi(x) \neq x$ for all x in X_n.

(a) Find a closed-form expression for D_n the number of derangements of n objects.

(b) Show that D_n is equal to the nearest integer to $\frac{n!}{e}$.

(c) Find the radius of convergence of the power series $\sum_{n=1}^{\infty} \frac{D_n x^n}{n!}$.

P1998-7. Let us say that two non-singular 2×2 matrices (with complex entries) A and B are *equivalent* if there exists a complex number ω such that $A = \omega B$. Let $[A]$ denote the equivalence class of A, that is, the set of all matrices equivalent to A. Note that multiplication of such equivalence classes is a well-defined operation.

(a) Find as simple a method as possible for calculating the square root(s) of an equivalence class of matrices other than $\begin{pmatrix} 1 & 0 \\ 0 & 1 \end{pmatrix}$ Here is an important example to consider: the square root(s) of $\begin{pmatrix} 2 & 5 \\ 1 & 3 \end{pmatrix}$ are

$$\begin{pmatrix} 3 & 5 \\ 1 & 4 \end{pmatrix} \quad \text{and} \quad \begin{pmatrix} 1 & 5 \\ 1 & 2 \end{pmatrix}$$

since

$$\begin{pmatrix} 3 & 5 \\ 1 & 4 \end{pmatrix}\begin{pmatrix} 3 & 5 \\ 1 & 4 \end{pmatrix} = \begin{pmatrix} 14 & 35 \\ 7 & 21 \end{pmatrix} = 7\begin{pmatrix} 2 & 5 \\ 1 & 3 \end{pmatrix}$$

and

$$\begin{pmatrix} 1 & 5 \\ 1 & 2 \end{pmatrix}\begin{pmatrix} 1 & 5 \\ 1 & 2 \end{pmatrix} = \begin{pmatrix} 6 & 15 \\ 3 & 9 \end{pmatrix} = 3\begin{pmatrix} 2 & 5 \\ 1 & 3 \end{pmatrix}.$$

(b) How many distinct square roots can an equivalence class that is not equal to $\begin{pmatrix} 1 & 0 \\ 0 & 1 \end{pmatrix}$ have? Realize that $[-A] = [A]$ so that the usual sign ambiguity doesn't exist here.

(c) How many distinct square roots does $\begin{pmatrix} 1 & 0 \\ 0 & 1 \end{pmatrix}$ have?

Exam #34–1999

This year the competition was held on the Indiana University in Bloomington. The team of Thomas Horine, James Lee, and Keith Henderson of Purdue University won the contest. Teams from Rose-Hulman and Wabash College placed second and third, respectively. The Rose-Hulman team consisted of Matt Lepinski, Dennis Lin, Randy Motchan, and the Wabash team consisted of Robert Dirks, Barry Weliver, Jun Ma.

P1999-1. In a large urn there are 1999 orange balls and 2000 yellow balls. Next to the urn is a large pile of yellow balls. The following procedure is performed repeatedly.

Two balls are chosen at random from the urn:

(i) If both are yellow, one is put back, the other thrown away;

(ii) If both are orange, they are both thrown away and a yellow ball from the pile is put into the urn;

(iii) if they are of different colors, the orange one is put back into the urn and the yellow one is thrown away.

What is the color of the last ball in the urn?

P1999-2. Let P be any point inside an equilateral triangle T. Show that the sum of the three distances from P to the sides of T is constant.

P1999-3. If x and y are positive numbers, show that

$$\sqrt{\frac{(\frac{x}{1})^2 + (\frac{x+y}{2})^2}{2}} \leq \frac{\sqrt{(\frac{x}{1})^2} + \sqrt{\frac{x^2+y^2}{2}}}{2}$$

and find all cases of equality.

P1999-4. The statement

$$\frac{1}{n}\sum_{k=1}^{n} f\left(\frac{k}{n}\right) \to \int_0^1 f(x)\,dx \qquad \text{as } n \to \infty$$

is familiar to all students of calculus. Several textbooks claim that the convergence is monotone in n. Show that this claim is false even for continuous increasing functions.

P1999-5. What is the probability that a dart, hitting a square board at random, lands nearer the center than the edge?

P1999-6. Does any row of Pascal's triangle contain three consecutive entries that are in the ratio $1:2:3$? Recall that the nth row ($n = 0, 1, 2, \ldots$) consists of the entries

$$\binom{n}{k} = \frac{n!}{k!(n-k)!} \qquad (k = 0, 1, 2, \ldots, n).$$

P1999-7. Does the series

$$\sum_{k=1}^{\infty} \frac{\cos(\ln(k))}{k}$$

converge or diverge?

P1999-8. Let A be a square matrix and suppose that there exist positive integers m and n such that $A^m = I$ and $A^n \neq I$. Calculate

$$\det(I + A + A^2 + \cdots + A^{m-1}).$$

Exam #35–2000

This year's winning team was from Purdue University and consisted of James Lee, Thomas Horine, and Keith Henderson. The second place team from Wabash College consisted of Roberts Dirks, Zhenyu Yang, and Chris Duefel. The team consisting of Matt Lepinski, Dennis Lin, and Lucas Beverlin from Rose-Hulman came in third. The contest was held at Earlham College.

P2000-1. Four suspects, one of whom was known to have committed a murder, made the following statements when questioned by police. If only one of them is telling the truth, who did it?

Arby: *Becky did it.*
Becky: *Ducky did it.*
Cindy: *I didn't do it.*
Ducky: *Becky is lying.*

P2000-2. Call a number N *fortunate* if it can be written with four equal digits in some base $b \in \mathbf{Z}^+$.

(a) Clearly 2222 is fortunate; why is 2000 fortunate?

(b) Find the greatest fortunate number less than 2000.

P2000-3. Determine whether

$$\sqrt{1 + \sqrt{2 + \sqrt{3 + \sqrt{4 + \sqrt{5 + \sqrt{6 + \cdots}}}}}}$$

converges or diverges.

P2000-4. Prove that a group G of order 15 must be cyclic.

P2000-5. Let the Fibonacci sequence F_k be defined as $F_0 = 0$, $F_1 = 1$, and $F_{k+2} = F_{k+1} + F_k$ for $k \geq 0$. It is known (and you may assume) that F_k is the closest integer to $\tau^k / \sqrt{5}$ where $\tau = (1 + \sqrt{5})/2$. Show that

$$\left(\sum_{k=0}^{\infty} \frac{(-1)^k F_k}{k!} \right) \left(\sum_{k=0}^{\infty} \frac{F_k}{k!} \right) = \frac{2}{5} \left(1 - \cosh \sqrt{5} \right).$$

P2000-6. In a triangle one angle is twice another. What is the largest possible ratio of the area of the triangle to that of its circumcircle? You may express your answer in terms of

$$g = \sqrt{\frac{4 + \sqrt{10}}{12}}.$$

P2000-7. Prove or disprove the statement: an irrational power of an irrational number is irrational. You may assume that $\sqrt{2}$ is irrational but be sure to prove all other results used.

P2000-8. Find all functions $f(x)$ which satisfy the equation

$$f(x) + f''(x) = f^{(3)}(x) + f^{(5)}(x)$$

and have

$$\lim_{x \to \infty} f(x) = f(0) = 0.$$

Solutions

Exam #1-1966

S1966-1

We will solve the more general problem of determining which integers A yield integer solutions x and y to the equation $x^2 - y^2 = A$. Since the LHS factors, we have the system of equations

$$x - y = r \quad \text{and} \quad x + y = s,$$

with the condition that $A = rs$. The system has solutions

$$x = \frac{r + s}{2} \quad \text{and} \quad y = \frac{r - s}{2}.$$

Thus x and y are integers whenever r and s have the same parity. This happens for all odd values of A ($r = A$, $s = 1$) and when A is a multiple of 4.

For our particular problem, let $a^3 = A$. If a is odd, then so is A, and if a is even, then A is a multiple of 4, so solutions exist for all positive integer values of a.

Look under Diophantine Equations in the Index for similar problems.

S1966-2

Divide the unit square into four smaller squares each of side length $1/2$ (by drawing two lines which bisect parallel sides of the original square). Since there are five points and only four smaller squares, two points must lie in the same square. It is clear that any two such points can be separated by a distance of less than $\sqrt{2}/2$ unless the two points are at opposite ends of the diagonal of the smaller square, which cannot happen since they are both interior to the original square.

Look under Geometry in the Index for similar problems.

S1966-3

Consider the numbers

$$0, a_1, a_1 + a_2, a_1 + a_2 + a_2, \ldots, a_1 + a_2 + \cdots + a_p.$$

There are $p + 1$ numbers in this list so two of them are congruent mod p. The difference between those two is the required sum divisible by p. (Note that p being prime is irrelevant.)

Look under Number Theory in the Index for similar problems.

S1966-4

Solution 1: Suppose the functions are $f(x)$ and $g(x)$, then we must have

$$\left(\frac{f(x)}{g(x)} \right)' = \frac{f'(x)}{g'(x)}.$$

Applying the quotient rule, you get

$$\frac{g(x) f'(x) - f(x) g'(x)}{(g(x))^2} = \frac{f'(x)}{g'(x)}.$$

Multiply this equation by $(g(x))^2 g'(x)$ to simplify the equation and we get

$$g(x) g'(x) f'(x) - (g'(x))^2 f(x) = (g(x))^2 f'(x).$$

If we know one of the functions, say $g(x)$, then we have a first order linear differential equation for the other function, that we can solve. Choosing $g(x) = x$ so $g'(x) = 1$, we find that $f(x)$ must satisfy

$$x f'(x) - f(x) = x^2 f'(x).$$

Rewriting this equation, we get

$$f'(x) = \frac{df}{dx} = \frac{1}{x(1-x)} f(x),$$

or

$$\frac{df}{f} = \frac{dx}{x(1-x)}.$$

Integrating, we have

$$\ln |f(x)| = \ln |x| - \ln |x - 1| + C,$$

so $f(x)$ must be
$$f(x) = C\,\frac{x}{x-1}.$$

An example of two functions $f(x)$ and $g(x)$ having the property that the derivatives of their quotient is the quotient of their derivatives is given by $f(x) = x/(x-1)$ and $g(x) = x$.

Solution 2: Begin the same as in Solution 1, but let $g(x) = \exp(kx)$, so that $g'(x) = ke^{kx}$ and the differential equation simplifies to
$$(k-1)f'(x) = f(x).$$

Solving this yields
$$f(x) = Ce^{\frac{x}{k-1}}.$$

Then, for any $k \neq 0$, 1, the functions f and g satisfy the necessary requirement.

Look under Differentiation or Real-Valued Functions in the Index for similar problems.

SI966-5

Solution 1: (In what follows, "sequence" refers to an ascending sequence of positive integers as in the problem.) Let $S(n,k)$ be the number of different sequences of length k in which every number from 1 to n occurs at least once. Then $S(n,k)$ is simply the number of ways to partition a k element sequence into n non-empty subsets. Hence, $S(n,k) = \binom{k-1}{n-1}$. Let $T(n,k,r)$ be the number of different sequences of length k in which only r (out of n) distinct numbers occur, then we have

$$T(n,k,r) = \binom{n}{r}S(r,k) = \binom{n}{r}\binom{k-1}{r-1}.$$

Finally, we see that the solution is

$$\sum_{r=1}^{n} T(n,k,r) = \sum_{r=1}^{n}\binom{n}{r}\binom{k-1}{r-1} = \binom{n+k-1}{k}$$

where the latter equality is an invocation of the Vandermonde convolution.

Solution 2: Make $n+k-1$ blanks and fill in k of them with x's. For any such arrangement of x's, define

$$a_i = 1 + \text{the number of blank to the left of the } i\text{th } x$$

for each i from 1 to k. This gives a 1-to-1 correspondence between the sequences we are trying to count and the ways of putting k x's in some of the $n + k - 1$ blanks. But the number of ways of doing the latter is obviously $\binom{n+k-1}{k}$.

Look under Enumeration in the Index for similar problems.

S1966-6

Solution 1: Let $f(x) = \sqrt{ax^2 + b}$. We want to show that $f(x)$ is a contraction, i.e. $\exists\, 0 < C < 1$ s.t. $|f(x) - f(y)| \leq C|x - y|$ for all $x, y \in \mathbf{R}$. In this case, the sequence $x_1 = c, x_{n+1} = f(x_n)$ converges to the unique fixed point of f.

$$
\begin{aligned}
|f(x) - f(y)| &= |\sqrt{ax^2 + b} - \sqrt{ay^2 + b}| \\
&= \sqrt{a}\,|\sqrt{x^2 + b/a} - \sqrt{y^2 + b/a}| \\
&\leq \sqrt{a}\,|x - y|.
\end{aligned}
$$

To see that $|\sqrt{x^2 + c} - \sqrt{y^2 + c}| \leq |x - y|$ for any $c \geq 0$, Multiply and divide by the conjugate to get

$$
\begin{aligned}
|\sqrt{x^2 + c} - \sqrt{y^2 + c}| &= \frac{|x^2 - y^2|}{|\sqrt{x^2 + c} + \sqrt{y^2 + c}|} \\
&= \frac{|x - y||x + y|}{|\sqrt{x^2 + c} + \sqrt{y^2 + c}|}.
\end{aligned}
$$

Since

$$
|\sqrt{x^2 + c} + \sqrt{y^2 + c}| \geq |x| + |y| \geq |x + y|,
$$

the inequality follows.

Since $0 < \sqrt{a} < 1$, f is a contraction. The limit of the sequence $\{x_n\}$ is simply the fixed point of f. Solving $x = \sqrt{ax^2 + b}$ yields $\lim_{n \to \infty} x_n = \sqrt{b/(1 - a)}$.

Solution 2: After computing the first several terms of the sequence, we see that

$$
\begin{aligned}
x_{n+1} &= \sqrt{a^n c^2 + b(1 + a + a^2 + \cdots + a^{n-1})} \\
&= \sqrt{a^n c^2 + b\left(\frac{1 - a^n}{1 - a}\right)}.
\end{aligned}
$$

Since $0 < a < 1$, we may take the limit giving

$$\lim_{n \to \infty} (x_{n+1}) = \sqrt{0 \cdot c^2 + b\left(\frac{1-0}{1-a}\right)} = \sqrt{\frac{b}{1-a}}.$$

Look under Limit Evaluation in the Index for similar problems.

S1966-7

Assume that the set $\{I_k\}$ is minimal in the sense that none of the intervals is a proper subset of the union of some others. Then we can order the intervals by the value of their left-endpoint, so that I_i begins to the left of I_{i+1} for all i. Then we have I_j and I_{j+2} are disjoint for each $j = 1, \dots, n-2$; otherwise $I_{j+1} \subset I_j \bigcup I_{j+2}$, or $I_{j+2} \subset I_{j+1}$, either of which violate the minimality assumption.

Thus, the set of intervals with odd subscripts consists of mutually disjoint sets, as does the set of intervals with even subscripts. At least one of these two sets covers $1/2$ of I, since their union covers all of I.

S1966-8

Choose a person P. If P is a friend of k other people, then P is a stranger of $5 - k$ other people, and one of k or $5 - k$ must be at least 3. Without loss of generality, assume that P is a friend of at least 3 other people: Q_1, Q_2, Q_3. If Q_1, Q_2, Q_3 are all pairwise strangers, then we are done. Otherwise, Q_i and Q_j are friends (for some i, j). In this case, P, Q_i, Q_j forms a triple of pairwise friends.

Look under Enumeration in the Index for similar problems.

Exam #2–1967

S1967-1

In a symmetric matrix, all non-diagonal entries are paired with another; the two must be the same. There are $n(n-1)$ such entries. If each row is a permutation of the set $\{1, 2, \dots, n\}$, then there must be n copies of each number in the matrix and, since n is odd, at most $n - 1$ copies of each number can paired off. The remaining copy of each number must be on the diagonal of the matrix, implying that the diagonal is also a permutation of the set $\{1, 2, \dots, n\}$.

Look under Matrix Algebra in the Index for similar problems.

S1967-2

Without loss of generality, assume that the axes of the parabolas lie parallel to the x-axis; let (x_1, y_1) and (x_2, y_2) be the points of contact of the common tangent with the parabolas. Then the equations for the parabolas can be given by

$$y^2 + 2B_1 x + 2C_1 y + D_1 = 0$$

and

$$y^2 + 2B_2 x + 2C_2 y + D_2 = 0.$$

Using implicit differentiation, we have the tangent line to parabola i through (x_i, y_i) is given by

$$y y_i + B_i(x + x_i) + C_i(y + y_i) + D_i = 0.$$

Since both of the points (x_1, y_1) and (x_2, y_2) lie on each of these lines, we have

$$y_1 y_2 + B_1(x_1 + x_2) + C_1(y_1 + y_2) + D_1 = 0 \qquad (1)$$

and

$$y_1 y_2 + B_2(x_1 + x_2) + C_2(y_1 + y_2) + D_2 = 0. \qquad (2)$$

Subtracting equation (2) from equation (1), we obtain

$$2(B_1 - B_2)\left(\frac{x_1 + x_2}{2}\right) + 2(C_1 - C_2)\left(\frac{y_1 + y_2}{2}\right) + (D_1 - D_2) = 0,$$

that is, the midpoint lies on the line

$$2(B_1 - B_2)x + 2(C_1 - C_2)y + (D_1 - D_2) = 0,$$

which is the common chord of the parabolas. (The common chord is the line segment connecting the two points of intersection of the parabolas. That this chord exists is implicitly assumed here.)

Look under Analytic Geometry in the Index for similar problems.

S1967-3

Let $f(a, b) = e^a + b(\log b - 1) - ab$, for $a \geq 0$ and $b \geq 1$. If $f(a, b) \geq 0$, then the result follows. Note that if $b = e^a$, then $f(a, b) = 0$.

Further, since

$$\frac{\partial f}{\partial b} = \log b - a,$$

we have

$$\frac{\partial f}{\partial b} = 0 \text{ if } b = e^a, \qquad \frac{\partial f}{\partial b} < 0 \text{ if } b < e^a,$$

and

$$\frac{\partial f}{\partial b} > 0 \text{ if } b > e^a.$$

Therefore, fixing a_0 greater than or equal to zero, consider the function $f(a_0, b)$, for $1 \le b \le \infty$. From the facts about $\partial f / \partial b$ above, $f(a_0, b)$ has an absolute minimum at $b = e^{a_0}$. Further, since $f(a_0, e^{a_0}) = 0$, we have $f(a_0, b) \ge 0$, for $1 \le b \le \infty$. Since a_0 was arbitrary, the result follows.

Look under Real-Valued Functions in the Index for similar problems.

S1967-4

Suppose the proposition false and let n be the smallest positive integer such that the number of odd coefficients in the expansion of $(x + y)^n$ is not a power of 2. By inspection n is not 1, nor is n equal to 2.

Case 1.   Let $n = 2k$, with k a positive integer. If m is odd, then

$$\binom{n}{m} = \frac{n - m + 1}{m} \binom{n}{m-1}$$

is an even integer since $n - m + 1$ is even. Thus all the odd coefficients must occur for even values of m. Let $m = 2q$. Then $\binom{n}{m}$ is congruent (mod 2) to

$$\frac{2 \times 4 \times \cdots \times n}{2 \times 4 \times \cdots \times m \times 2 \times 4 \times \cdots \times (n - m)}$$

by removing all odd factors and divisors. (That is, if x and y are congruent (mod 2) then so are hx and ky for any odd integers h and k, and conversely.) But the latter fraction is evidently $\binom{k}{q}$, thus $\binom{n}{m} = \binom{2k}{2q}$ is odd if, and only if, $\binom{k}{q}$ is odd and so the number of odd coefficients in $(x + y)^n$ is the same as the number of odd coeficients in $(x + y)^k$, which contradicts the assumption that n was the least positive integer for which the proposition fails.

Case 2.   Let $n = 2k + 1$, with k a positive integer. Since $\binom{n}{m} = \binom{n}{n-m}$ and, since n and $n - m$ are not congruent (mod 2), the number of odd coefficients is twice the number of odd coefficients obtained by considering only even values of m in the symbols $\binom{n}{m}$. But as before we find $\binom{2k+1}{2q} \equiv \binom{k}{q}$ (mod 2). Thus the number of odd coefficients in $(x + y)^n$, is just twice the number of odd coefficients in $(x + y)^k$, again giving a contradiction. Thus the proposition is true.

Look under Enumeration in the Index for similar problems.

SI967-5

Suppose that z is a root and $|z| < 1/3$. Then

$$1 + a_1 z + a_2 z^2 + \cdots + a_N z^N = 0$$

implies that

$$a_1 z + a_2 z^2 + \cdots + a_N z^N = -1$$

or

$$|a_1 z + a_2 z^2 + \cdots + a_N z^N| = 1.$$

But using the triangle inequality, the requirement that $|a_n| < 2$, and $|z| < 1/3$, we obtain

$$|a_1 z + a_2 z^2 + \cdots + a_N z^N| < 2\left(\frac{1}{3} + \frac{1}{3^2} + \cdots + \frac{1}{3^N}\right) = 1 - \frac{1}{3^N} < 1,$$

a contradiction.

Look under Complex Numbers or Polynomials in the Index for similar problems.

SI967-6

In the illustration below, the distance from A to F, and the distance from B to G are each $\sqrt{65}/8$. The distances AG, AH, AT, BH, and BJ are all larger. If the square is divided as illustrated, each piece has a diameter of $\sqrt{65}/8$.

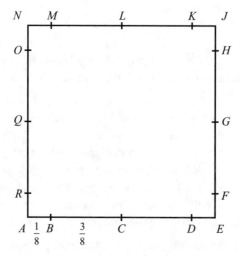

Now suppose that the square is divided into three arbitrary pieces. One of the three pieces must contain at least six of the sixteen boundary points symmetrically arranged in the illustration. We will now show that this piece has a diameter of at least $\sqrt{65}/8$.

Suppose the set contains a corner point, WLOG point A. If the set contains any of points F, G, H, J, K, L, or M, the set has diameter at least $\sqrt{65}/8$. If the set contains none of those points, but contains, WLOG, corner E then the set must also contain one of N, P, Q, or R, and again the diameter is at least $\sqrt{65}/8$. Finally, if the set contains none of E, F, G, H, J, K, L, M, or N, then the set must contain one or both of D and P, and one or both of C and Q. Again the diameter is at least $\sqrt{65}/8$.

Suppose the set contains none of the corner points, but does contain a non-midpoint of one of the sides, WLOG, say B. If the set also contains G, H, K, or L, the set has diameter at least $\sqrt{65}/8$. If the set contains none of those points, but does contain M, then it must contain one of C, D, or F. Again the set has diameter at least $\sqrt{65}/8$. If the set doesn't contain M either, it may contain P. In this case, it must also contain D or F, and hence has a diameter of at least $\sqrt{65}/8$. This leaves the possibility of the set consisting of B, C, D, F, Q, R, in which case it has a diameter of at least $\sqrt{65}/8$.

This leaves the possibility of a set consisting solely of the midpoints, but there are only four, so this can't happen.

Look under Geometry in the Index for similar problems.

S1967-7

First note that for $a > 0, y > 0$, the equation $x + x^a = y$ has a unique solution with $x > 0$. This is because $x + x^a$ is a continuous, monotonic increasing function of x with $\lim_{x \to \infty} x + x^a = \infty$ and is equal to zero at $x = 0$. Thus it assumes every positive real value exactly once for $x \in (0, \infty)$. Now, if the sequence $\{x_n\}$ satisfies the conditions given in the problem, then

$$x_n = \sum_{j=n+1}^{\infty} x_j^a = x_n + 1^a + \sum_{j=n+2}^{\infty} x_j^a$$
$$= x_{n+1}^a + x_{n+1}.$$

So x_{n+1} must be the unique solution of this equation. If follows that the sequence $\{x_n\}$ is unique. Further, if we define x_n inductively by setting

x_{n+1} equal to the solution of $x + x^a = x_n$, then it is clear that

$$x_n = x_{n+1}^a + x_{n+1}$$

$$= x_{n+1}^a + x_{n+2}^a + x_{n+1} = \cdots$$

$$= \sum_{j=n+1}^{\infty} x_j^a.$$

Look under Infinite Series in the Index for similar problems.

S1967-8

Suppose that $p, q \in \mathbf{R}^2$ and let m be the midpoint of p and q. Let S be the line segment of length 1 which is perpendicular to $\overline{pq}$ and has m as its midpoint. By symmetry, for every $s \in S$, $|p - s| = |q - s|$. Now, for each $n \in N$, let $s_n \in S$ be such that $|p - s_n| \in \mathbf{Q}$ and $|m - s_n| < 1/n$. Such a choice is possible because $\mathbf{Q}$ is dense in $\mathbf{R}$ and $|p - x|$ is a continuous function of x. So as x varies along S in a $\frac{1}{n}$-neighborhood of m, $|p - x|$ assumes rational values. Now, as $n \to \infty$, $s_n \to m$, hence $|p - s_n| + |q - s_n| \to |p - q|$, and we see that

$$|T(p) - T(q)| \le |T(p) - T(s_n)| + |T(q) - T(s_n)|$$

$$= |p - s_n| + |q - s_n|.$$

Letting $n \to \infty$, we have $|T(p) - T(q)| \le |p - q|$.

Now, suppose $\varepsilon > 0$. We can choose $q' \in \mathbf{R}^2$ such that $|q - q'| < \varepsilon$ and $|p - q'| \in Q$. As a consequence of the previous argument, we have that $|T(q) - T(q')| \le |q - q'|$. Also, we can invoke the triangle inequality to see that $\left| |T(p) - T(q)| - |T(p) - T(q')| \right| \le |T(q') - T(q)| \le |q - q'| < \varepsilon$.

$$\left| |T(p) - T(q)| - |p - q| \right|$$

$$\le \left| |T(p) - T(q)| - |p - q'| \right| + \left| |p - q'| - |p - q| \right|$$

$$= \left| |T(p) - T(q)| - |T(p) - T(q')| \right| + \left| |p - q'| - |p - q| \right|$$

$$\le \left| |T(p) - T(q)| - |T(p) - T(q')| \right| + |q - q'|$$

$$< 2\varepsilon.$$

Letting $\varepsilon \to 0$, we see that $|T(p) - T(q)| = |p - q|$ for all $p, q \in \mathbf{R}^2$.

Look under Geometry in the Index for similar problems.

Exam #3–1968

S1968-1

Since the set of Riemann sums is bounded, there exists an M such that

$$M \geq \Big| \sum_{i=1}^{n} f(t_i)(x_i - x_{i-1}) \Big| \quad \text{for all } t_i, x_i, x_{i-1}.$$

Thus we have

$$M \geq |f(t_i)(b - a)| \quad \text{for all } t_i$$

implying

$$M/(b - a) \geq |f(t_i)| \quad \text{for all } t_i,$$

that is, that f is bounded.

Look under Riemann Sums in the Index for similar problems.

S1968-2

Assume the claim is false for quadrilateral $ABCD$ and five interior points. If P and Q are two of the five interior points, then the line through P and Q cannot intersect $ABCD$ in adjacent sides. To see this, suppose, without loss of generality, that the line intersects AB and BC. Then the points A, C, D, P, and Q form a convex pentagon. Therefore, exactly one of the rays $\overrightarrow{PQ}$ or $\overrightarrow{QP}$ intersects either AB or BC. Let S_{AB} be those rays $\overrightarrow{PQ}$ intersecting AB and S_{BC} be those rays $\overrightarrow{PQ}$ intersecting BC, where P and Q range over the five interior points. Thus $S_{AB} \cup S_{BC}$ contains ten rays, one for each pair of points P, Q. If rays $\overrightarrow{PQ}$ and $\overrightarrow{QR}$ are both in, say, S_{AB}, then either the points P, Q, R, B, and C or the points P, Q, R, A, and D form a convex pentagon, depending on which side of line PR the point Q lies. It follows furthermore that if $\overrightarrow{PQ}$ and $\overrightarrow{PR}$ or if $\overrightarrow{QP}$ and $\overrightarrow{RP}$ are both in, say, S_{AB}, then $\overrightarrow{QR}$ or $\overrightarrow{RQ}$ must be in S_{BC}. Without loss of generality, assume S_{AB} contains at least five rays. Because no point can be both the initial and second point of such rays, it follows that some point P is the initial or second point in at least three of these rays. Let the other points of these rays be Q, R, and S. The previous discussion showed that the rays $\overrightarrow{QR}$ or $\overrightarrow{RQ}$, $\overrightarrow{QS}$ or $\overrightarrow{SQ}$, and $\overrightarrow{RS}$ or $\overrightarrow{SR}$ must be in S_{BC}. However, this possibility is also precluded by the previous discussion, proving the claim.

Look under Geometry in the Index for similar problems.

SI968-3

Fix t and define $L : \mathbf{R}^n \mapsto \mathbf{R}^n$ by $L(x) = tx$. Then L is differentiable on $\mathbf{R}^n$, with $L'(x) = tI$, where I is the identity matrix. Now let $F(x) = f(L(x))$, so that by the chain rule we have $F'(x) = f'(L(x))L'(x) = tf(tx)$. By the definition of f, $F(x) = tf(x)$, so that $F'(x) = tf'(x)$ also. Therefore $f'(tx) = f'(x)$ for any $t > 0$. Therefore $f'(x)$ is constant and $f(x)$ must be linear.

Look under Multivariate Calculus in the Index for similar problems.

SI968-4

The only integral solutions to the first equation are $(1,0)$ and $(2,1)$. The only solutions to the second are $(2,3)$ and $(1,1)$. This is a special case of Catalan's Conjecture and a general exposition of the progress that has been made on the conjecture can be found in MAA FOCUS, vol 21 #5.

Look under Diophantine Equations in the Index for similar problems.

SI968-5

This conjecture is false. Let

$$f_n(x) = \begin{cases} \frac{1}{nx}, & \text{on } [1, e^n], \\ 0, & \text{otherwise.} \end{cases}$$

Since $0 \leq f_n(x) \leq \frac{1}{n}$, for all x, $f_n \to 0$ uniformly on $\mathbf{R}$. Now

$$\int_{-\infty}^{\infty} f_n(x)\, dx = 1$$

so that $\lim_{n \to \infty} \int_{-\infty}^{\infty} f_n = 1$.

Look under Infinite Series, Limit Evaluation or Riemann Sums in the Index for similar problems.

SI968-6

Consider any two sequences $\{c_n\}$ and $\{d_n\}$ each tending monotonically to zero, with $d_n > c_n$, and $\sum c_n$ converging while $\sum d_n$ diverges.

Since $\sum d_n$ diverges, for each n there exists an integer $\phi(n)$ such that $\sum_{j=n}^{\phi(n)} d_j \geq e$ where e is any fixed positive number. Define the sequence n_k recursively by $n_1 = 1$ and $d_{n_{k+1}} < c_{\phi(n_k)}$. It is clear that $n_k < \phi(n_k) < n_{k+1}$.

Now a solution to the problem can be given by the two series with terms

$$\{a_n\} = \{c_{n_1}, c_{n_1+1}, \ldots, c_{\phi(n_1)}, d_{n_2}, d_{n_2+1}, \ldots, d_{\phi(n_2)}, c_{n_3}, \ldots,$$

$$c_{\phi(n_3)}, d_{n_4}, \ldots\},$$

$$\{b_n\} = \{d_{n_1}, d_{n_1+1}, \ldots, d_{\phi(n_1)}, c_{n_2}, c_{n_2+1}, \ldots, c_{\phi(n_2)}, d_{n_3}, \ldots,$$

$$d_{\phi(n_3)}, c_{n_4}, \ldots\}.$$

Clearly, each of the series $\sum a_n$ and $\sum b_n$ diverge since they contain infinitely many stretches of terms adding up to more than e, and $\sum \min(a_n, b_n)$ converges by comparison with $\sum c_n$.

Look under Infinite Series in the Index for similar problems.

SI968-7

$f(z, \theta) = \max\{|z| \cos(\theta - \arg z), 0\}$. Then

$$\frac{1}{2\pi} \int_{-\pi}^{\pi} f(z, \theta)\, d\theta = \frac{|z|}{2\pi} \int_{-\pi/2}^{\pi/2} \cos \theta\, d\theta = \frac{|z|}{\pi}.$$

Therefore,

$$\frac{1}{2\pi} \int_{-\pi}^{\pi} \sum_{j=1}^{n} f(z_j, \theta)\, d\theta = \sum_{j=1}^{n} \frac{|z_j|}{\pi} = \frac{1}{\pi}.$$

Because the maximum of a function is at least its average, it follows that there exists θ such that

$$\frac{1}{\pi} \leq \sum_{j=1}^{n} f(z_j, \theta)$$

$$= \sum_{\cos(\theta - \arg z_j) \geq 0} |z_j| \cos(\theta - \arg z_j)$$

$$= \sum_{\cos(\theta - \arg z_j) \geq 0} |z_j| \cos(\arg z_j - \theta)$$

$$= Re \sum_{\cos(\theta - \arg z_j) \geq 0} z_j e^{-i\theta} \leq \left| \sum_{\cos(\theta - \arg z_j) \geq 0} z_j e^{-i\theta} \right|$$

$$= \left| \sum_{\cos(\theta - \arg z_j) \geq 0} z_j \right|.$$

Look under Complex Numbers in the Index for similar problems.

S1968-8

This problem asks you to look for the Ramsey number $n = r(k, k)$. We will give an existence result for the more general problem of finding $r(s, t)$, that is, for what n can we find a set of s blue arcs or a set of t arcs. In particular, we show that

$$r(s, t) \le r(s - t, t) + r(s, t - 1).$$

Working inductively on $k = s + t$, we have the result holding for $k = 4$ by inspection. Now suppose the result holds for all values smaller than k.

Let $n = r(s - 1, t) + r(s, t - 1)$ be the number of points that we have, and let each arc be colored blue or red. Let v be any point. There are either $r(s - 1, t)$ blue arcs or $r(s, t - 1)$ red arcs at v. If there are $r(s - 1, t)$ blue arcs, then the set of points at the other end of these arcs contain $s - 1$ blue arcs or t red arcs among them, by our induction hypothesis. In the first case these blue arcs, plus those from v to these points, form a collection of s blue arcs, as required. In the second case, we already have t points all of whose arcs are red.

A similar argument holds if there are $r(s, t - 1)$ red arcs at v.

Look under Enumeration in the Index for similar problems.

Exam #4–1969

S1969-1

We will show that

$$f(x) = \sin x - x + \frac{x^2}{\pi} \ge 0$$

on $[0, \pi]$. By observation of $f'(x) = \cos x - 1 + \frac{2x}{\pi}$, we find that $0, \frac{\pi}{2}$, and π are critical points of $f(x)$. By noting that

$$f''(x) = -\sin x + \frac{2}{\pi}$$

has exactly two zeros on $[0, \pi]$, we can apply Rolle's Theorem to verify that $0, \frac{\pi}{2}$, and π are the only critical points of $f(x)$. (If f' had another zero, f'' would have a third zero.) Since $f(0) = f(\pi) = 0$ and $f(\frac{\pi}{2}) = 1 - \frac{\pi}{4}$, we can see that the minimum value of $f(x)$ on $[0, \pi]$ is 0.

Look under Real-Valued Functions in the Index for similar problems.

S1969-2

Suppose that

$$\frac{P}{p} + \frac{Q}{q} + \frac{R}{r} = A.$$

Multiply through by pqr yielding $Pqr + Qpr + Rpq = Apqr$. Now,

$$Pqr = Apqr - Qpr - Rpq = p(Aqr - Qr - Rq),$$

hence p divides Pqr. But since $(p,q) = 1$ and $(p,r) = 1$, it must be that p divides P. In the same manner, q divides Q and r divides R.

Look under Number Theory in the Index for similar problems.

S1969-3

Let $\omega = \sqrt[3]{25 + 5\sqrt{20}}$ and $\bar{\omega} = \sqrt[3]{25 - 5\sqrt{20}}$. We wish to determine $s = \omega + \bar{\omega}$. We have

$$s^3 = \omega^3 + \bar{\omega}^3 + 3\omega\bar{\omega}(\omega + \bar{\omega}) = 50 + 3 \cdot 5 \cdot s.$$

To solve for s we note that $0 = s^3 - 15 - 50 = (s-5)(s^2 + 5s + 10)$ has one real solution, $s = 5$.

Look under Number Theory in the Index for similar problems.

S1969-4

Label acceleration, a, downward at all points along the trajectory of height h, and g the acceleration due solely to gravity. Let v be the velocity of the ball at any time, in the general direction of travel (downward or upward) and v_0 be the upward velocity with which the ball is launched.

At launch, $a = g + \alpha v$, so v continually decreases until the highest point of ascent, where $v = 0$ (by Rolle's theorem), so $a = g$. Thus, going upward, the ball will always experience an acceleration downward greater than g. The ascent will take the same amount of time as an object travelling on a straight line with initial velocity $v = 0$ and initial acceleration $a = g$, with a continually increasing through the distance h, to a final value of $a = g + \alpha v_0$. Since a is continually increasing, the amount of time to traverse this distance, t_{up}, is less than t, the amount of time for an object with uniform acceleration g to traverse this distance.

Going downward, the acceleration begins with $a = g$ and continues with $a = g - \alpha v$ (since the air resistance is now upward, but the velocity is downward). So the descent will take the same amount of time as an object travelling on a straight line with initial acceleration $a = g$, with a continually decreasing through the distance h to a final value of $a = g - \alpha v_f$. Again, t_{down}, the time required to traverse this distance, is greater than t, the time an object under uniform acceleration g would require. Thus, $t_{up} < t < t_{down}$.

Look under Differential Equations in the Index for similar problems.

S1969-5

Let θ denote the angle just above A. Using the law of sines to compute the left side of the equilateral triangle, we find its area to be

$$\frac{\sqrt{3}}{4}\left[\frac{b\sin(2\pi/3 - \theta)}{\sqrt{3}/2} + \frac{c\sin(\theta + A - \pi/3)}{\sqrt{3}/2}\right]^2$$

$$= \frac{1}{\sqrt{3}}\left[b\sin(2\pi/3 - \theta) - c\sin\left((2\pi/3 - \theta) - (A + \pi/3)\right)\right]^2$$

$$= \frac{1}{\sqrt{3}}\left[\left(b\sin(2\pi/3 - \theta) - c\cos(A + \pi/3)\right)\sin(2\pi/3 - \theta)\right.$$

$$\left. + c\sin(A + \pi/3)\cos(2\pi/3 - \theta)\right]^2.$$

Now, if ω is such that $\cos\omega = x/\sqrt{x^2 + y^2}$ and $\sin\omega = y/\sqrt{x^2 + y^2}$, we may write $x\sin\alpha + y\cos\alpha$ as $\sqrt{x^2 + y^2}\sin(\alpha + \omega)$, from which we see that its maximum value is $\sqrt{x^2 + y^2}$. It follows that the maximum area of the equilateral triangle is

$$\frac{\sqrt{3}}{4}\left[\left(b - c\cos(A + \pi/3)\right)^2 + \left(c\sin(A + \pi/3)\right)^2\right]$$

$$= \frac{\sqrt{3}}{4}\left(b^2 + c^2 - 2bc\cos(A + \pi/3)\right).$$

The claimed upper bound has been established. However, it need not be the maximum value. To conform to the geometric setup, θ must satisfy $0 < \theta < 2\pi/3$ and $0 < \pi - (\theta + A) < 2\pi/3$, which is equivalent to $\pi/3 - A < \theta < \pi - A$. For instance, for $b = 10$, $c = 1$, and $A = \pi - .1$, the maximum occurs for θ outside this range.

Look under Geometry in the Index for similar problems.

S1969-6

From the Trapezoidal Rule we have

$$\int_a^b f(x)\,dx = \frac{b-a}{2n}\left(f(a) + 2\sum_{k=1}^{n-1} f(a+kh) + f(b)\right) - \frac{(b-a)^3}{12n^3}f''(c)$$

for some $c \in [a,b]$. If we set $n = 1$, the result follows immediately. A complete proof of the Trapezoidal Rule itself can be found many places.

Look under Real-Valued Functions in the Index for similar problems.

Exam #5–1970

S1970-1

We have

$$\lim_{n\to\infty}\left(\frac{1}{n+1} + \frac{1}{n+2} + \cdots + \frac{1}{2n}\right) = \lim_{n\to\infty}\sum_{k=1}^{n}\frac{1}{n+k}$$

$$= \lim_{n\to\infty}\sum_{k=1}^{n}\left(\frac{1}{n}\cdot\frac{1}{\frac{k}{n}+1}\right)$$

$$= \int_0^1 \frac{1}{1+x}\,dx = \ln(2).$$

Look under Limit Evaluation in the Index for similar problems.

S1970-2

If $|c| \le 1$, then we are done. If $|c| > 1$, then we have

$$c^n = -a_{n-1}c^{n-1} - a_{n-2}c^{n-2} - \cdots - a_1 c - a_0.$$

Divide by c^{n-1} to obtain

$$c = -a_{n-1} - a_{n-2}c^{-1} - \cdots - a_1 c^{-n+2} - a_0 c^{-n+1}.$$

Take absolute values and use the triangle inequality to see that

$$|c| \le |a_{n-1}| + |a_{n-2}/c| + \cdots |a_1/c^{n-2}| + |a_0/c^{n-1}|.$$

Since $|c| > 1$, we have the desired inequality.

Look under Polynomials in the Index for similar problems.

SI970-3

Solution 1:   Without loss of generality, we may assume that the coordinates of the vertices of the parallelogram are $A: (0,0)$, $B: (1,0)$, $C: (a,b)$, $D: (a+1, b+1)$. Then the coordinates of X, Y, Z, W are

$$X: \left(\frac{1}{2}, -\frac{1}{2}\right), \quad Y: \left(1 + \frac{a+b}{2}, \frac{b-a}{2}\right),$$

$$Z: \left(a + \frac{1}{2}, b + \frac{1}{2}\right), \quad W: \left(\frac{a-b}{2}, \frac{a+b}{2}\right).$$

Then $\overline{XZ} = \sqrt{a^2 + (b+1)^2} = \overline{YW}$.

We also have $\vec{XZ} = \langle a, b+1 \rangle$ and $\vec{YW} = \langle b+1, -a \rangle$. Thus, $\vec{XZ} \cdot \vec{YW} = 0$ and $\vec{XZ} \perp \vec{YW}$.

Since the diagonals of a parallelogram bisect each other, the length of each side is the hypotenuse of an isosceles right triangle with legs of length $\sqrt{a^2 + (b+1)^2}/2$ and $XYZW$ must be a rhombus.

To see that the rhombus is a square, let E denote the intersection of XZ with YW. Then $m\angle WXY = m\angle WXE + m\angle YXE =$ a right angle. Similarly, the other angles of $XYZW$ are right angles, and $XYZW$ is a square.

Solution 2:   Let YW and XZ meet at P. Then P is a center of symmetry for the figure, so a $180°$ rotation maps the figure to itself. Hence $WXYZ$ is a parallelogram. Next, $ZC = ZD$ and $CY = DW$. Also $\angle CZD = 90°$, $\angle ZCB = \angle ZDW = 90° +$ the acute angle of the given parallelogram. Hence triangles ZCY and ZDW are congruent and similarly oriented, so a $90°$ rotation carries one triangle into the other. Thus $ZY = ZW$ and $\angle YZW = 90°$. Thus $WXYZ$ is a square.

Look under Geometry in the Index for similar problems.

SI970-4

The problem is trivial for $e = 2$, so we proceed by induction. Suppose the result is true for all sets of points of order e, where $e = k$, k even. Consider a set of points and arcs of order $e = k + 2$. Pick any arc A, say from P_1 to P_2 and remove it together with its endpoints. By our induction hypothesis, we can color the remaining regions with two colors.

Because of the non-intersection constraint on the arcs, P_1 and P_2 must either be consecutive points around the circle, or be situated such that arc A divides the remaining arcs into two sets, as illustrated below. In the first case, color the region created by arc A opposite of the color of the region it is contained in. In the second case, color one of the two regions created by arc A the other color (from what is was before) and switch colors for all of the other regions on that side of arc A.

Look under Geometry in the Index for similar problems.

S1970-5

Let d be the number of digits in n, then it is necessary that $6 \cdot 10^{d-1} + (n-6)/10 = 4n$ (from condition (b)). Multiplying through by 10 yields $6 \cdot 10^d + (n-6) = 40n$ (note here that n will also satisfy (a)) or equivalently

$$n = \frac{6 \cdot 10^d - 6}{39} = \frac{2(10^d - 1)}{13}.$$

Hence it is necessary that 13 divide $10^d - 1$. So we need

$$10^d - 1 = 9 \sum_{k=0}^{d-1} 10^k \equiv 0 \,(\mathrm{mod}\ 13).$$

Since $\gcd(9, 13) = 1$, this is equivalent to

$$\sum_{k=0}^{d-1} 10^k \equiv \sum_{k=0}^{d-1}(-3)^k \equiv 0 \,(\mathrm{mod}\ 13).$$

Now we may simply add alternating powers of 3 until we reach zero: $(-3)^0 = 1, 1 + (-3)^1 = -2, -2 + (-3)^2 = 7, 7 + (-3)^3 = -20 = -7, -7 + (-3)^4 = 74 = 9, 9 + (-3)^5 = 9 - 243 = -234 = -18 \cdot 13 = 0$. So $d - 1 = 5 \Rightarrow d = 6$. Thus if $n = 2(10^6 - 1)/13 = 153, 846$ satisfies (a) and (b) then it is the smallest such number. Indeed, since d is the correct number of digits, (a) and (b) are true by construction.

Exam #6-1971

S1971-1

For part (a) we have

(i) $x \in A \Rightarrow x \in S \Rightarrow xRx$ (with $x \in A$) $\Rightarrow x \in \overline{A}$, hence $A \subset \overline{A}$.

(ii) By (1), $\overline{A} \subset \overline{\overline{A}}$. Now, $x \in \overline{\overline{A}} \Rightarrow \exists y \in \overline{A}$ s.t. xPy. $y \in \overline{A} \Rightarrow \exists z \in A$ s.t. yPz. $xPy \wedge yPz \Rightarrow xPz$ (with $z \in A$), so $x \in \overline{A}$. It follows that $\overline{\overline{A}} \subset \overline{A}$, hence $\overline{A} = \overline{\overline{A}}$.

(iii) $x \in \overline{A \cup B} \Rightarrow \exists y \in A \cup B$ s.t. xPy. If $y \in A$, then $x \in \overline{A}$ and if $y \in B$, then $x \in \overline{B}$. Therefore $\overline{A \cup B} \subset \overline{A} \cup \overline{B}$. Also, $x \in \overline{A} \cup \overline{B} \Rightarrow (x \in \overline{A}) \vee (x \in \overline{B})$. So one of A or B contains a y s.t. xPy. It follows that $x \in \overline{A \cup B}$ and hence $\overline{A \cup B} = \overline{A} \cup \overline{B}$.

(b) $\overline{A} = \{(u,v) \in S : v - y = 3(u - x) \text{ for some } (x,y) \in A\} = \{(u,v) \in S : v - 3u = y - 3x \wedge x^2 + y^2 = 1\}$. First, we determine the values that $y - 3x$ can take when $x^2 + y^2 = 1$. Thus we look at the minima and maxima of the two functions $\sqrt{1 - x^2} - 3x$ and $-\sqrt{1 - x^2} - 3x$ on $[-1, 1]$. Since these functions are continuous, they will assume every value between their respective minimum and maximum.

$$\frac{d}{dx}\left(\pm\sqrt{1 - x^2} - 3x\right) = \mp\frac{x}{\sqrt{1 - x^2}} - 3 = 0 \Rightarrow 9(1 - x^2)$$

$$= x^2 \Rightarrow x = \pm\frac{3}{\sqrt{10}}.$$

Checking the values $x = \pm 1, \pm\frac{3}{\sqrt{10}}$ in $\pm\sqrt{1 - x^2} - 3x$ yields a minimum and maximum of $-\sqrt{10}$ and $\sqrt{10}$, respectively. Hence $y - 3x$ assumes every value on $[-\sqrt{10}, \sqrt{10}]$. Now we can write

$$\overline{A} = \{(u,v) \in S : v - 3u \in [-\sqrt{10}, \sqrt{10}]\}$$

$$= \{(u,v) \in S : 3u - \sqrt{10} \le v \le 3u + \sqrt{10}\}.$$

It follows that $\overline{A}$ is the strip between the two parallel lines $v = 3u - \sqrt{10}$ and $v = 3u + \sqrt{10}$.

Look under Algebraic Structures in the Index for similar problems.

S1971-2

We want to compute $7^{9999} \pmod{1000}$.

$$7^{9999} \equiv 7^{-1}(7^{400})^{25} \equiv 7^{-1}(10^3 k + 1)^{25} \equiv 7^{-1} \cdot 1 \equiv 143 \pmod{1000}.$$

So the last three digits are 1, 4, 3.

Look under Number Theory in the Index for similar problems.

S1971-3

The derivative of $(1 - e^{\frac{-1}{x}})^{-1}$ is

$$\frac{e^{\frac{-1}{x}}}{x^2(1 - e^{\frac{-1}{x}})^2}.$$

Note, however that this is valid only for $x \neq 0$; worse yet, $(1 - e^{\frac{-1}{x}})^{-1}$ has a jump discontinuity at $x = 0$.

(b) $(1 - e^{\frac{-1}{x}})^{-1}$ is not an antiderivative of the integrand on the interval $[-1, 1]$, but the function g is, where

$$g(x) = \begin{cases} (1 - e^{\frac{-1}{x}})^{-1} & \text{for } x > 0, \\ 1 & \text{for } x = 0, \\ 1 + (1 - e^{\frac{-1}{x}})^{-1} & \text{for } x < 0. \end{cases}$$

Hence the integral is $g(1) - g(-1) = \frac{2}{e-1}$.

A more direct approach would be to express the integral as the sum of two integrals, one on the interval $[-1, 0]$ and the other on the interval $[0, 1]$.

Look under Differentiation or Integration in the Index for similar problems.

S1971-4

For each $t \in \mathbf{R}$, let L_t be the line $y(x) = t$. Clearly the set $B = \{L_t : t \in \mathbf{R}\}$ is uncountable, $L_{t_1} \cap L_{t_2} = \emptyset$ for $t_1 \neq t_2$, and $\bigcup_{t \in \mathbf{R}} L_t = \mathbf{R}^2$. It follows from the latter two statements that for any $a \in A$, there is exactly one line L_t for which $a \in L_t$. Let $f : A \to B$ be the function which takes points in A to the line in B which contains them. Since A is countable and B is uncountable, f cannot be surjective. It follows that there exists an $L_t \in B$ which does not contain a point of A.

Look under Enumeration in the Index for similar problems.

S1971-5

For each positive integer k,

$$(2 + \sqrt{2})^k + (2 - \sqrt{2})^k = \sum_{j=0}^{\lfloor k/2 \rfloor} \binom{k}{2j} 2^{k-j+1},$$

which is an integer; but $0 < (2 - \sqrt{2})^k < 1$. Therefore, the fractional part of of $(2 + \sqrt{2})^k$ is $1 - (2 - \sqrt{2})^k$, and its limit is 1.

(b) A similar calculation shows that the fractional part of

$$(1 + \sqrt{2})^k = \begin{cases} (\sqrt{2} - 1)^k & \text{if } k \text{ is odd}, \\ 1 - (\sqrt{2} - 1)^k & \text{if } k \text{ is even}. \end{cases}$$

The first of these has limit 0 and the second the limit 1. Hence the fractional part of of $(1 + \sqrt{2})^k$ has cluster points at 0 and 1.

Look under Limit Evaluation in the Index for similar problems.

S1971-6

For each group of five robbers, there must be at least one lock to which none has the key, and for any two groups, these locks must be different (else the addition to one group of a new robber from the other would provide a majority no member of which contains a key for such a lock as we have associated with the first group). Therefore there are at least as many locks as there are groups of five robbers, namely $\binom{11}{5} = 462$.

Now consider any one robber. For each group of five among the remaining 10 robbers, he must have a key to each lock to which they collectively do not. Since each group has at least one such lock, he must have at least as many keys as there are such groups, namely $\binom{10}{5} = 252$.

In order to show that no more locks or keys are required, we now demonstrate that 462 locks and 252 keys are sufficient. In accordance with any one-to-one correspondence between a set of 462 different locks and the collection of six-member subsets of the robber band, we give keys for any one lock to each member of its associated subset and to no other robber. Each minority, being a subset of the complement of such a six-member subset, cannot open the lock associated with that subset. On the other hand, consider any majority and any lock. Since only five robbers do not have the key to that lock, at least one of the robbers in the given majority does have the key.

Look under Enumeration in the Index for similar problems.

S1971-7

We define three new items from the given operations as follows:

$$\bigcup = *(*A \bigcap *B),$$

$$\emptyset = X \bigcap *X, \quad \text{for all } X, \text{ and}$$

$$A \subset B \quad \text{iff } A \bigcap B = A.$$

With these definitions and the axioms given we can show, after a considerable amount of simple but lengthy work, that we do indeed have a Boolean Algebra.

In fact, the converse is also true. If you start with the standard axioms for a Boolean Algebra, you can derive (a), (b), and (c). Therefore, the conditions of the problem give an alternate characterization of a Boolean Algebra.

Look under Algebraic Structures in the Index for similar problems.

Exam #7–1972

S1972-1

The following calculation shows that it is sufficient that $A + B$ and $A - B$ be invertible. Supposing the existence of X and Y such that the desired conditions hold, the equations

$$AX + BY = C,$$

$$BX + AY = D$$

yield $(A + B)(X + Y) = C + D$. Subtracting the second from the first yields $(A - B)(X - Y) = C - D$. If $A + B$ and $A - B$ are invertible, then

$$X - Y = (A - B)^{-1}(C - D),$$

$$X + Y = (A + B)^{-1}(C + D),$$

and from these equations X and Y can be found. The only conditions are that A, B, C, and D are the same size.

Look under Matrix Algebra in the Index for similar problems.

S1972-2

From the given equation

$$I = A\left(-\frac{1}{2}A^2 - 2A - \frac{3}{2}I\right),$$

so A^{-1} exists and is equal to $-\frac{1}{2}A^2 - 2A - \frac{3}{2}I$.

S1972-3

Let $f(x) = \ln x / x$. Then $f'(x) = (1 - \ln x)/x^2$ and $f'(x) = 0$ when $\ln x = 1$, or when $x = e$. Since $f'(x)$ changes sign from $+$ to $-$ as x passes through e, f has a relative maximum at $x = e$. Further, it is an absolute maximum since $f(x) \to 0$ as $x \to \infty$, and $f(x) \to -\infty$ as $x \to 0$. Thus $f(\pi) < f(e)$, so $\ln \pi/\pi < \ln e/e$ whence $\pi^e < e^\pi$.

Look under Matrix Algebra or Real-Valued Functions in the Index for similar problems.

S1972-4

There exists N such that for all $n > N$, sufficiently large, $0 < a_n < 1/2$. Thus for $n > N$, $a_n/(1 - a_n) < 2a_n$. So

$$0 < \sum_{n=1}^{\infty} \frac{a_n}{1 - a_n} < \sum_{n=1}^{N} \frac{a_n}{1 - a_n} + 2 \sum_{n=N+1}^{\infty} a_n$$

and the series converges.

Look under Infinite Series in the Index for similar problems.

S1972-5

Let P denote the permutation made by the machine. We are given

$$P^2 = (A, 10, J, 6, 3, Q, 2, 9, 5, K, 7, 4, 8)$$

in cycle notation. Since P^2 is a cyclic permutation, P is also. Therefore P^{13} is the identity permutation. Thus

$$P = (P^2)^7 = (A, 9, 10, 5, J, K, 6, 7, 3, 4, Q, 8, 2).$$

Look under Permutations in the Index for similar problems.

S1972-6

The curves are identical, since (r, θ) and $(-r, 3\pi - \theta)$ are different coordinates for the same point. If $r = \cos(\theta/2)$, then

$$-r = \cos \frac{3\pi - \theta}{2} = \cos \frac{3}{2}\pi \cos(\theta/2) + \sin \frac{3}{2}\pi \sin \theta/2 = -\sin \theta/2.$$

S1972-7

Suppose $G = H \cup K$ with H and K proper subgroups. Since H is proper, there exists $h \in H$ such that $h \notin K$. Similarly, there exists $k \in K$ such that $k \notin H$. Let $g = hk$. If $g \in H$, then $k = h^{-1}g \in H$, which is impossible; if $g \in K$, then $h = gk^{-1} \in K$, which is also impossible. Thus G cannot be $H \cup K$.

Look under Group Theory in the Index for similar problems.

Exam #8–1973

S1973-1

Using L'Hopital's rule,

$$\lim_{x\to\infty} x(e^{1/x} - 1) = \lim_{x\to\infty} \frac{e^{1/x} - 1}{1/x}$$

$$= \lim_{x\to\infty} \frac{e^{1/x}(-1/x^2)}{-1/x^2} = \lim_{x\to\infty} e^{1/x} = 1.$$

Look under Limit Evaluation in the Index for similar problems.

S1973-2

Let α be the angle shown in in the figure below, with α variable and θ fixed.

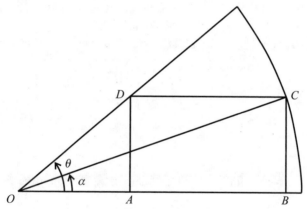

Let A be the area of rectangle $ABCD$. Since $\cot\theta = \frac{OA}{AD}$ and $BC = OC \sin\alpha = \sin\alpha$, We have

$$AB = OB - OA$$
$$= OC\cos\alpha - AD\cot\theta$$
$$= \cos\alpha - BC\cot\theta$$
$$= \cos\alpha - \sin\alpha\cot\theta.$$

Thus the area, A, of the rectangle is given by

$$A = BC \cdot AB$$
$$= \sin\alpha\cos\alpha - \sin^2\alpha\cot\theta$$
$$= \frac{1}{2}\sin 2\alpha - \sin^2\alpha\cot\theta,$$

which we differentiate to obtain

$$\frac{dA}{dr} = \cos 2\alpha - \sin 2\alpha \cot \theta.$$

Setting this equal to 0, we find $\alpha = \theta/2$. Substituting into our area function, we get

$$A = \frac{1 - \cos \theta}{2 \sin \theta}.$$

Look under Max/Min Problems in the Index for similar problems.

S1973-3

Observe that the maximum occurs when no $n_i = 1$. Further, $(m-2)\cdot 2 > m$ if, and only if, $m > 4$. Thus, each $n_i > 4$ can be replaced by $(n_i - 2) + 2$ in the partition and the associated product will be increased. If some $n_i = 4$, replace it by $2 + 2$, leaving the product unchanged. If three 2's occur replace them by $3 + 3$, increasing the product. Thus the maximum occurs when each n_i is a 2 or a 3, with at most two 2's occurring. Indeed, the maximum is:

$$\frac{n-1}{3^3} \cdot 2^{i/2}, \quad \text{where} \quad \begin{cases} i = 4 & \text{if } n \equiv 1 \ (\mathrm{mod}\ 3), \\ i = 2 & \text{if } n \equiv 2 \ (\mathrm{mod}\ 3), \\ i = 0 & \text{if } n \equiv 0 \ (\mathrm{mod}\ 3). \end{cases}$$

S1973-4

Integration by parts with $u = 1/(1 + y)$ and $dv = y^{n-1} dy$ gives

$$\int_0^1 \frac{n y^{n-1}}{1 + y} \, dy = \frac{y^n}{1 + y} \Big|_0^1 + \int_0^1 \frac{y^n}{(1 + y)^2} \, dy.$$

Since

$$0 \le \int_0^1 \frac{y^n}{(1 + y)^2} \, dy \le \int_0^1 y^n \, dy = 1/(n + 1),$$

the integral on the right-hand side of the last equation approaches zero as n approaches infinity. The desired limit is thus

$$\lim_{n \to \infty} \left(\frac{y^n}{1 + y} \Big|_0^1 \right) = 1/2.$$

Look under Integration or Limit Evaluation in the Index for similar problems.

S1973-5

The quotient $A : B$ is an ideal of R. If $a \in A : B$, then $\forall b \in B$,

$$ab \in A \Rightarrow -(ab) \in A \Rightarrow (-a)b \in A,$$

hence $-a \in A : B$. If $x, y \in A : B$, then

$$\forall b \in B, \quad xb, yb \in A \Rightarrow xb + yb \in A \Rightarrow (x + y)b \in A,$$

hence $x + y \in A : B$. Finally, if $a \in A : B, r \in R$, then $\forall b \in B$, $ab \in A \Rightarrow r(ab) \in A \Rightarrow (ra)b \in A$ so that $ra \in A : B$ (in particular, $0 \in A : B$). It follows that $A : B$ is an ideal.

S1973-6

Let $f(a) = a^{n+1} - n(a - 1) - a$. Then

$$\frac{d}{da} f(a) = (n + 1)a^n - (n + 1).$$

Since $f(1) = f'(1) = 0$, $f(a)$ is divisible by $(a - 1)^2$.

Look under Polynomials in the Index for similar problems.

S1973-7

We will prove the contrapositive: If A or B is singular, then the product AB is singular. Let B be singular. Then for some $x \neq 0$, we have $Bx = O$. It follows that

$$A(Bx) = (AB)x = O$$

implying that AB is singular. Now let A be singular. Then for some $x \neq 0$, $Ax = O$. If B is non-singular, there is a non-trivial solution to $By = x$. Hence

$$Ax = A(By) = (AB)y = O,$$

from which we see that AB is singular.

Look under Matrix Algebra in the Index for similar problems.

S1973-8

Let $x = a - d$, $y = d$, and $z = a + d$. Then

$$(a - d)^2 + d^2 = (a + d)^2,$$

or

$$a^2 - 2ad + d^2 + d^2 = a^2 + 2ad + d^2,$$

or

$$a^2 = 4ad.$$

Thus $a = 0$ or $a = 4d$, and all solutions are $(x, y, z) = (-d, 0, d)$ or $(3d, 4d, 5d)$, where d is any integer.

Look under Number Theory in the Index for similar problems.

S1973-9

There are $\binom{5}{3} = 10$ equiprobable ways to choose 3 men from the group. Exactly 3 of those choices result in 2 Republicans being chosen (2 Republicans and the choice of 1 out of 3 Democrats). It follows that the probability of both Republicans being selected is $3/10$.

Look under Probability in the Index for similar problems.

Exam #9–1974

S1974-1

(a) For a polynomial $p(x)$, let $[x^m]p(x)$ denote the coefficient of x^m in $p(x)$. First, by the binomial theorem,

$$(x + 1)^n = \sum_{k=0}^{n} \binom{n}{k} x^k.$$

From this, we see that

$$[x^m](x + 1)^n = \binom{n}{m}.$$

It follows that

$$[x^m](x^{1000-k}(x + 1)^k) = [x^{m-(1000-k)}](x + 1)^k$$
$$= \binom{k}{m - (1000 - k)} = \binom{k}{1000 - m}.$$

Now, the polynomial given in the problem is equal to

$$\sum_{k=0}^{1000} x^{1000-k}(x + 1)^k,$$

so

$$[x^m] \sum_{k=0}^{1000} x^{1000-k}(x+1)^k = \sum_{k=0}^{1000} [x^m](x^{1000-k}(x+1)^k)$$

$$= \sum_{k=0}^{1000} \binom{k}{1000-m}$$

$$= \binom{1001}{1001-m} = \binom{1001}{m}.$$

Therefore the coefficient of x^{50} is $\binom{1001}{50}$.

(b) From part (a), we see that the sum of all the coefficients is

$$\sum_{m=0}^{1000} [x^m] = \sum_{m=0}^{1000} \binom{1001}{m} = 2^{1001} - 1.$$

Look under Polynomials in the Index for similar problems.

S1974-2

We have

$$yx = (yx)^3 = (yx)(yx)^2 = (yx)^2(yx)$$

$$= ((yx)^2 y)x = y(yx)^2 x = y(yx)(yx)x$$

$$= xy^2yx^2 = xyx^2 = x^3 y = xy.$$

Look under Group Theory in the Index for similar problems.

S1974-3

If $m > 1$, set

$$n_1 = m - 1, \quad N_1 = m^2 - m + 1, \quad n_2 = m + 1, \quad N_2 = m^2 + m + 1.$$

Then we have

$$(m^2 + 1)(n_1^2 + 1) = (m^2 + 1)((m-1)^2 + 1)$$

$$= (m^2 + 1)(m^2 - 2m + 2)$$

$$= (m^4 - 2m^3 + 3m^2 - 2m + 1) + 1$$

$$= (m^2 - m + 1)^2 + 1$$

$$= N_1^2 + 1,$$

$$(m^2 + 1)(n_2^2 + 1) = (m^2 + 1)\big((m+1)^2 + 1\big)$$
$$= (m^2 + 1)(m^2 + 2m + 2)$$
$$= (m^4 + 2m^3 + 3m^2 + 2m + 1) + 1$$
$$= (m^2 + m + 1)^2 + 1$$
$$= N_2^2 + 1.$$

If $m = 1$, then $n_1 = 2, N_1 = 3, n_2 = 12, N_2 = 17$ are the smallest solutions.

S1974-4

The point (0,0) is neither a minimum nor a maximum. We have $f(0,0) = 0$; let $x = \delta$ and $y = \sqrt{\delta} - \epsilon$ where $0 < \epsilon < \delta/2$. Then $f(x,y) < 0$. But $f(x,0) > 0$ for all x.

Look under Multivariate Calculus in the Index for similar problems.

S1974-5

Assume that y is never zero on $[0, 3/4]$. We note that $\frac{dy}{dx} \le -1$, so $y(3/4) > 0$ means that $y(x) > \frac{3}{4} - x$ on $[0, 3/4]$. Therefore

$$\frac{dy}{dx} \le -1 - e^x y \le -1 - e^x\left(\frac{3}{4} - x\right).$$

We see that

$$y(3/4) = 1 + \int_0^{3/4} \frac{dy}{dx}\, dx < 1 - \int_0^{3/4} 1 + e^x\left(\frac{3}{4} - x\right)\, dx = 2 - e^{3/4}.$$

Since $\ln(2) \approx .7$, $2 - e^{3/4} < 0$.
    (Or, $2^{4/3} = 16^{1/3} < 17.576^{1/3} = 2.6 < e$.)
    Thus, $y(3/4) < 0$, contradicting our assumption. Hence, y must have a zero on $[0, 3/4]$.

Look under Real-Valued Functions in the Index for similar problems.

S1974-6

This conjecture is clearly false. Let $a_n = 1/(n+1)$, then

$$\sum_{n=0}^{\infty} a_n = \sum_{n=1}^{\infty} \frac{1}{n}$$

(the harmonic series) diverges while

$$\sum_{n=0}^{\infty} a_n^2 = \sum_{n=1}^{\infty} \frac{1}{n^2} = \frac{\pi^2}{6}.$$

Look under Infinite Series in the Index for similar problems.

Exam #10–1975

S1975-1

The result is obvious if $a = 0$, so consider $a > 0$. Also let

$$f(x) = \frac{a}{a^2 + x^2}.$$

We see, using a lower Riemann Sum, that

$$0 < \sum_{n=1}^{\infty} \frac{a}{a^2 + n^2} < \int_o^{\infty} \frac{a}{a^2 + x^2} \, dx = \frac{\pi}{2}$$

and the result follows. If $a < 0$, the infinite sum is an odd function of a, so the result follows in this case also.

Look under Infinite Series in the Index for similar problems.

S1975-2

First we show, by way of contradiction, that no two of the vertices can be joined by a diameter through the center of the circle. Let V be the set of vertices, with $|V| = n$, odd. Suppose that v_0 and v_0' are two vertices joined by a chord through the center of the circle. This chord partitions $V - \{v_0, v_0'\}$ into two disjoints sets, V_1 and V_2, of orders a and b respectively. We have $a + b = n - 2$, and without loss of generality, a is odd and b is even.

Label the vertices of V_1 as $v_1^1, v_1^2, \ldots, v_1^a$ as we move around the circle from v_0 to v_0'. Label the vertices of V_2 as $v_2^1, v_2^2, \ldots, v_2^b$ as we again move around the circle the other way from v_0 to v_0'.

Now each of the $\frac{a+1}{2}$ triplets

$$(v_0, v_1^1, v_1^2), (v_1^2, v_1^3, v_1^4), \ldots, (v_1^{a-1}, v_1^a, v_0'),$$

has the same angle, say β. Also note that the angles

$$\angle(v_0, O, v_1^2), \angle(v_1^2, O, v_1^4), \ldots, \angle(v_1^{a-1}, O, v_0'),$$

(where O is the center of the circle) are all equal to $2\pi - 2\beta$, and they sum to π. Therefore a simple calculation gives us $(a+1)(\pi - \beta) = \pi$.

Now consider the $\frac{b}{2}$ triplets

$$(v_0, V_2^1, v_2^2), (v_2^2, v_2^3, v_2^4), \ldots, (v_2^{b-2}, v_2^{b-1}, v_2^b).$$

As before, each of these form an angle of measure β. The angles

$$\angle(v_0, O, v_2^2), \angle(v_2^2, O, v_2^4), \ldots, \angle(v_2^{b-1}, O, v_2^b),$$

each have measure $2(\pi - \beta)$ and angle $\angle(v_2^b, O, v_0') = \alpha < 2(\pi - \beta)$. Therefore we have

$$\left(\frac{b}{2}\right)(2)(\pi - \beta) + \alpha = \pi,$$

$$(a+1)(\pi - \beta) = b(\pi - \beta) + \alpha,$$

and

$$\alpha = (\pi - \beta)(a - b + 1).$$

Now $0 < \alpha < \pi - \beta$ implies that $0 < a - b + 1 < 2$ and therefore, since a and b are integers, $a - b = 0$, a contradiction.

Therefore, a chord starting a vertex v_0 going through the center O of the circle and ending at the point w on the circle divides the other vertices into two sets of equal order. Thus

$$|V_1| = |V_2| = \frac{n-1}{2}.$$

If $\frac{n-1}{2}$ is even, we can proceed as before to obtain

$$\frac{1}{2}\frac{n-1}{2}(2)(\pi - \beta + \alpha_1) = \frac{1}{2}\frac{n-1}{2}(2)(\pi - \beta + \alpha_2)$$

as illustrated below. This implies that the triangles $O, w', v_1^{\frac{n-1}{2}}$ and $O, v', v_2^{\frac{n-1}{2}}$ are congruent. Letting the radius of the circle be r, this gives us the result that an arbitrary side of the polygon has length $2r \sin \alpha$, where

$$\alpha = \pi - \frac{n-1}{2}(\pi - \beta).$$

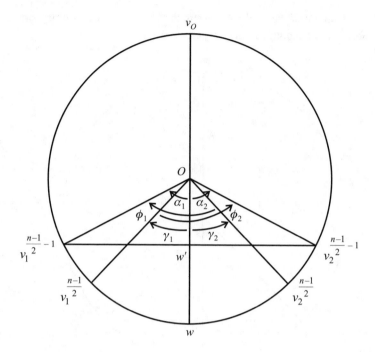

If $\frac{n-1}{2}$ is odd, we let

$$\angle(v_1^{\frac{n-1}{2}-1}, 0, w) = \alpha_1,$$

$$\angle(v_2^{\frac{n-1}{2}-1}, 0, w) = \alpha_1,$$

$$\angle(v_1^{\frac{n-1}{2}}, 0, w) = \gamma_1,$$

$$\angle(v_2^{\frac{n-1}{2}}, 0, w) = \gamma_2,$$

$$\angle(v_1^{\frac{n-1}{2}-1}, 0, v_2^{\frac{n-1}{2}}) = \phi_1,$$

$$\angle(v_2^{\frac{n-1}{2}-1}, 0, v_1^{\frac{n-1}{2}}) = \phi_1,$$

and can proceed as before to obtain

$$\frac{1}{2}\left(\frac{n-1}{2}-1\right)(2)(\pi - \beta + \alpha_1) = \frac{1}{2}\left(\frac{n-1}{2}-1\right)(2)(\pi - \beta + \alpha_2)$$

as illustrated below. This implies that the triangles $O, w', v_1^{\frac{n-1}{2}-1}$ and $O, w', v_2^{\frac{n-1}{2}-1}$ are congruent. From this we can see that $\gamma_1 = \gamma_2$, so that triangles $O, w', v_1^{\frac{n-1}{2}}$ and $O, w', v_2^{\frac{n-1}{2}}$ are congruent. Again letting the radius of the circle be r, this gives us the result that an arbitrary side of the polygon has length $2r \sin \gamma$, where $\gamma = \phi - \alpha$.

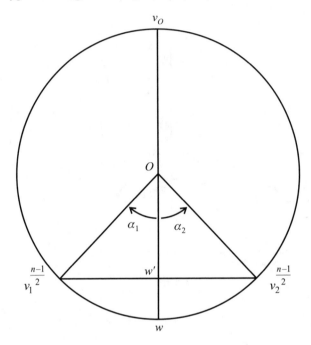

Look under Geometry in the Index for similar problems.

SI975-3

Let T_m be the number of components yielded by m lines in the plane. Clearly we have $T_0 = 1$ (the whole plane). Also, if there are $m - 1$ such lines forming T_{m-1} components, then adding an mth line according to the rules given will yield m more components. To see this, note that the new line will intersect each of the existing lines at a unique point. If there are $m - 1$ lines, then the new line will split m previous components yielding m new components. It follows that $T_m = T_{m-1} + m$. We see that the T_m are just partial sums so that

$$T_m = T_0 + \sum_{k=1}^{m} k = 1 + \sum_{k=1}^{m} k = 1 + \frac{m(m+1)}{2}.$$

Look under Enumeration in the Index for similar problems.

S1975-4

(a) If m is odd, then $3^m \equiv (-1)^m \equiv -1 \pmod 4$, so $3^m - 1 \equiv 2 \pmod 4$ and is not divisible by 2^m for odd m larger than 1.

(b) We note that $3^1 - 1 = 2$ is divisible by 2^1, $3^2 - 1 = 8 = 4 \cdot 2$ is divisible by 2^2, and $3^4 - 1 = 80 = 16 \cdot 5$ is divisible by 2^4.

By Euler's generalization of Fermat's little theorem $3^{\phi(2^m)} \equiv 1 \pmod{2^m}$, where $\phi(2^m)$ is the Euler totient function. Thus, $\phi(2^m) = 2^{m-1}$. Therefore the order of 3 modulo 2^m, $\mathrm{ord}_{2^m}(3)$, is a power of 2. Hence, $2^m | 3^m - 1$ only if m is a multiple of $\mathrm{ord}_{2^m}(3)$.

We now prove by induction that for $m \ge 4$, $3^{2^{m-3}} \equiv 1 + 2^{m-1} \pmod{2^m}$. When $m = 4$, we have $3^{2^1} = 9 \equiv 1 + 2^3 \pmod{2^4}$. Assume that the result is true for $m = k$, i.e., that $3^{2^{k-3}} \equiv 1 + 2^{k-1} \pmod{2^k}$.

Now consider the case $m = k + 1$. We recall that, by our induction hypothesis,

$$3^{2^{k-3}} \equiv 1 + 2^{k-1} + j2^k \pmod{2^{k+1}}.$$

Squaring, we see that

$$3^{2^{k-2}} \equiv 1 + 2^{2k-2} + j^2 2^{2k} + 2 \cdot 2^{k-1} + 2j2^k + 2j2^{2k-1} \pmod{2^{k+1}}.$$

Since $k \ge 4$, this is $3^{2^{k-2}} \equiv 1 + \cdot 2^k \pmod{2^{k+1}}$. Thus,

$$3^{2^{m-3}} \equiv 1 + 2^{m-1} \pmod{2^m}$$

and the claim is proved.

Next, we will prove that when $m \ge 3$, $\mathrm{ord}_{2^m}(3) = 2^{m-2}$. Since the order of 3 modulo 2^m is a power of two, we need only check exponents that are powers of 2. Since

$$3^{2^{m-3}} \equiv 1 + 2^{m-1} \pmod{2^m}, \quad \mathop{\mathrm{ord}}_{2^m}(3) > 2^{m-3}.$$

Squaring $3^{2^{m-3}}$, we see that

$$3^{2^{m-2}} \equiv 1 + 2 \cdot 2^{m-1} + 2^{2m-2} \pmod{2^m}.$$

As $m > 2$, we have $3^{2^{m-2}} \equiv 1 \pmod{2^m}$. Hence $\mathrm{ord}_{2^m}(3) = 2^{m-2}$.

When $m > 4$, $m < 2^{m-2}$. (This is also easy to see by induction. It's clear that $m = 5 < 8 = 2^3$ and $m+1 < m+m = 2m < 2 \cdot 2^{m-2} = 2^{m-1}$.) Therefore, $m \le 4$ and the only values for which $2^m | (3^m - 1)$ are $m = 1$, 2, or 4.

Look under Number Theory in the Index for similar problems.

S1975-5

By reversing the limits of integration, we have

$$\int_0^a g(t)dt = \int_0^a \int_x^a \frac{f(t)}{t} \, dt \, dx$$

$$= \int_0^a \int_0^t \frac{f(t)}{t} \, dx \, dt$$

$$= \int_0^a \frac{f(t)}{t} \cdot t \, dt = \int_0^a f(t) \, dt.$$

Look under Multivariate Calculus in the Index for similar problems.

Exam #11–1976

S1976-1

By inspection, $p(x) = x$ satisfies the conditions. Suppose that there is another solution, and let its degree be n. From $p(2) = 2$ follows $p(2^2 - 1) = (p(2))^2 - 1$ or $p(3) = 3$. Similarly, $p(8) = 8$ and $p(r) = r$ implies $p(r^2 - 1) = r^2 - 1$. Thus $n+1$ numbers a_1, $i = 1, 2, \ldots n+1$ can be found such that $p(a_i) = a_i$. Thus the polynomial $p(x) - x$, of degree n has $n + 1$ roots and this is impossible.

Look under Polynomials in the Index for similar problems.

S1976-2

We are given that $n \equiv 11 \pmod{12}$, whence $n \equiv 5 \pmod 6$. This implies that the prime-power decomposition of n contains a prime p, $p \equiv 5 \pmod 6$ raised to an odd power. (If not, it would follow that $n \equiv 1 \pmod 6$.) So we can write $n = p^e N$, $(p, N) = 1$, e odd. Then $\sigma(n) = \sigma(p^e)\sigma(N)$, and

$$\sigma(p^e) = 1 + p + p^2 + \cdots + p^{e-1} + p^e$$

$$\equiv 1 - 1 + 1 - \cdots + 1 - 1 \equiv 0 \pmod 6.$$

Hence $6|\sigma(n)$. If $n \equiv 11 \pmod{12}$, then $n \equiv 3 \pmod 4$, and the prime-power decomposition of n contains a prime q, $q \equiv 3 \pmod 4$, raised to an odd power. A calculation similar to the above shows that $4|\sigma(n)$, which gives the result.

Look under Number Theory in the Index for similar problems.

S1976-3

The quantity whose absolute value is to be shown to be less than 1 is $f'(0)$. Since

$$|f'(0)| = \left| \lim_{h \to 0} \frac{f(h)}{h} \right| = \lim_{h \to 0} \frac{|f(h)|}{|h|} \leq \lim_{h \to 0} \frac{|\tan h|}{|h|} = 1,$$

we are done.

Look under Differentiation or Real-Valued Functions in the Index for similar problems.

S1976-4

$f''' > 0$ for all x, so f'' has at most one zero, f' has at most two, and f at most three. Since $f(0) = f(1) = 0$, and $f'(1)3 \ln 3 - 4 < 3.3 - 4 < 0$, and $f(x) \to \infty$ as $x \to \infty$, f has another zero to the right of 1. f has exactly three zeros.

Look under Real-Valued Functions in the Index for similar problems.

S1976-5

Suppose that the equation of the circle is

$$x^2 + y^2 + rx + sy + t = 0.$$

Then

$$ra_1 + sa_2 + t = -a_1^2 - a_2^2,$$
$$rb_1 + sb_2 + t = -b_1^2 - b_2^2,$$
$$rc_1 + sc_2 + t = -c_1^2 - c_2^2.$$

The determinant of the coefficients is $(b_1 - a_1)(c_2 - b_2) - (c_1 - b_1)(b_2 - a_2)$, and since the points are not collinear

$$\frac{b_1 - a_1}{b_2 - a_2} \neq \frac{c_1 - b_1}{c_2 - b_2}.$$

Thus the determinant is not zero, and the system has a unique solution.

Look under Analytic Geometry in the Index for similar problems.

S1976-6

Since for all $x \in \mathbf{R}$, $x = 2y$ for some y, we have $f(x) = f(2y) = f(y)^2 \geq 0$. Further, $f(1) = f(0 + 1) = f(0)f(1)$, and since $f(1) \neq 0$,

$f(0) = 1$. Now let $L = \lim_{x \to 0} f(x)$. Then

$$1 = f(0) = f(x - x) = f(x)f(-x) \to L^2.$$

Since L cannot equal -1 since this would imply $f(x) \le 0$ for some x, we have $L = 1$.

Look under Limit Evaluation in the Index for similar problems.

Exam #12–1977

S1977-1

$$f(a) = f(b) \Rightarrow f(f(a)) = f(f(b)) \Rightarrow a = b;$$

therefore f is injective on J. Since f is continuous, it assumes all intermediate values. It follows that if f were not strictly monotone then it could not be injective.

Look under Real-Valued Functions in the Index for similar problems.

S1977-2

(a) Since, for $n = 1$, we have $\cos x = \frac{1}{2}(\cos x + \cos(-x))$, we proceed by induction. We therefore assume that

$$\cos^m x = \frac{1}{2^m} \sum_{k=0}^{m} \binom{m}{k} \cos(m - 2k)x$$

and consider

$\cos^{m+1} x$

$$= \frac{1}{2^m} \sum_{k=0}^{m} \binom{m}{k} \cos(m - 2k)x \cos x$$

$$= \frac{1}{2^{m+1}} \sum_{k=0}^{m} \left[\binom{m}{k} \cos\big((m + 1) - 2k\big)x + \cos\big((m - 1) - 2k\big)x \right]$$

since

$$\cos a \cos b = \frac{\cos(a + b) + \cos(a - b)}{2}.$$

The result now follows after some tedious, but routine manipulation of the summation, including shifting an index of summation and adding binomial coefficients.

(b) We have

$$\frac{1}{\pi} \int_0^\pi \cos^{2n} x \, dx = \frac{1}{\pi} \frac{1}{2^{2n}} \sum_{k=0}^{2n} \binom{2n}{k} \int_0^\pi \cos\left((2n-2k)x\right) dx$$

$$= \frac{1}{\pi} \frac{1}{2^{2n}} \left(\sum_{k=0, k\neq n}^{2n} \binom{2n}{k} \frac{\sin(2n-2k)x}{2n-2k} \bigg|_0^\pi + \binom{2n}{n}\pi \right)$$

$$= \frac{1}{\pi} \frac{1}{2^{2n}} \left(\frac{(2n)!}{(n!)^2}\pi \right) = \frac{(2n)!}{2^{2n}(n!)^2}.$$

Look under Integration or Trigonometry in the Index for similar problems.

S1977-3

Multiply the matrix A by the vector $\mathbf{1} = (1, 1, \dots, 1)$ to obtain $A\mathbf{1} = \mathbf{1}$. Thus $(A - I)\mathbf{x} = \mathbf{0}$ has a nonzero solution and $A - I$ is therefore not invertible.

Look under Matrix Algebra in the Index for similar problems.

S1977-4

Suppose that $f(0) = t$. Then

$$|f(t) - t| = |f(t) - f(0)| = |t|$$

and

$$|f(-t) - t| = |f(-t) - f(0)| = |t|.$$

Also, $|f(t) - f(-t)| = 2|t|$ so that one of $f(t) = 0$ or $f(-t) = 0$ must hold. It follows that $f(x) = 0$ for some x in **R**. Now for any y in **R**,

$$|f(y)| = |f(y) - f(x)| = |y - x|$$

so that any isometry must be of the form $f(y) = \pm(y - x)$ for some x in **R**. It is also easy to see that anything of that form must be an isometry. (These are translations followed by reflections.)

Look under Real-Valued Functions in the Index for similar problems.

S1977-5

Assume that there are only N primes. Then for each k, the kth partial sum of the harmonic series can be written

$$\sum_{n=1}^{k} \frac{1}{n} = \frac{1}{p_1^{a_1} p_2^{a_2} \cdots p_N^{a_N}} + \frac{1}{p_1^{b_1} p_2^{b_2} \cdots p_N^{b_N}} + \cdots + \frac{1}{p_1^{c_1} p_2^{c_2} \cdots p_N^{c_N}}$$

since each integer n can be written as a product of our N primes. From the hint, we therefore have

$$\sum_{n=1}^{k} \frac{1}{n} \leq \left(1 + \frac{1}{p_1} + \frac{1}{p_1^2} + \cdots + \frac{1}{p_1^r}\right)\left(1 + \frac{1}{p_2} + \frac{1}{p_2^2} + \cdots + \frac{1}{p_2^r}\right)$$

$$\cdots \left(1 + \frac{1}{p_N} + \frac{1}{p_N^2} + \cdots + \frac{1}{p_N^r}\right)$$

$$\leq \prod_{i=1}^{N} \frac{1}{1 - p_i}.$$

This implies that the harmonic sequence is increasing and bounded above, and therefore converging; a contradiction. (The hint is verified by picking $r \geq \max\{a_i, b_i, \ldots, c_i\}$, a finite set.)

Look under Number Theory in the Index for similar problems.

Exam #13–1978

S1978-1

To find the decimal expansion of an integer M, one divides by 10 successively, recording the remainders:

$$M = 10M_1 + d_0,$$

$$M_1 = 10M_2 + d_1,$$

$$\vdots$$

$$M_n = d_n,$$

where the d_i are $0, 1, \cdots, 8$, or 9 to obtain the expansion

$$M = d_n 10^n + \cdots + d_1 10 + d_0.$$

To write M in the base -2, one proceeds by an analogous algorithm, bearing in mind that the remainders must be 0 or $+1$. So

$$-2374_{(10)} = 101111001110_{(-2)}.$$

Look under Number Theory in the Index for similar problems.

S1978-2

To solve this problem, one must get at the area of a non-right-angled triangle. One technique is to use the cross product. Recall that $\|\mathbf{v} \times \mathbf{w}\|$ gives the area of the parallelogram with sides $\mathbf{v}$ and $\mathbf{w}$. To set this up, we may assume that A is the origin in 3-space, and that B, C, and D lie on the axes, say at $(x, 0, 0)$, $(0, y, 0)$, and $(0, 0, z)$, respectively. Then

$$\text{area}(ABC) = \frac{xy}{2}, \quad \text{area}(ABD) = \frac{xz}{2}, \quad \text{area}(ACD) = \frac{yz}{2},$$

and the area of (BCD) is

$$\frac{1}{2}\|\mathbf{CD} \times \mathbf{CB}\| = \frac{1}{2}\sqrt{(xy)^2 + (xz)^2 + (yz)^2}.$$

Look under Analytic Geometry in the Index for similar problems.

S1978-3

We recall that when k is odd, $a^k + b^k$ factors as

$$a^k + b^k = (a+b)(a^{k-1} - a^{k-2}b + a^{k-3}b^2 + \cdots + b^{k-1}).$$

Now $S(n) = \sum_{j=1}^{n} j^k$, and also $S(n) = \sum_{j=1}^{n}(n+1-j)^k$. Adding these two expressions, we get

$$2S(n) = \sum_{j=1}^{n}[(n+1-j)^k + j^k]$$

$$= \sum_{j=1}^{n}\left[(n+1-j+j)\sum_{i=0}^{k-1}(n+1-j)^{k-1-j}j^i\right]$$

$$= (n+1)\sum_{j=1}^{n}\left[\sum_{i=0}^{k-1}(n+1-j)^{k-1-j}j^i\right].$$

Look under Finite Sums in the Index for similar problems.

S1978-4

A first step in discovering the solution of this problem might be to graph the function for small values of n, as illustrated. The x for which $f(x)$ is smallest is the median of $a_1, a_2, \ldots, a_n$. That is, if n is odd, $f(x)$ is smallest at the middle a and if n is even, $f(x)$ is smallest in the interval between the middle a's.

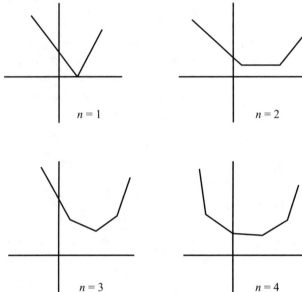

$n = 1$ $n = 2$

$n = 3$ $n = 4$

In view of this, the proof will be easier to write down if we re-index the given numbers so that $a_1 \leq a_2 \leq \cdots \leq a_n$. We divide the proof into two cases.

If $n = 2m + 1$, then

$$f(a_{m+1}) = \sum_{K=1}^{2m+1} |a_{m+1} - a_k|$$

$$= \sum_{k=1}^{m} (a_{m+1} - a_k) + \sum_{k=m+2}^{2m+1} (a_k - a_{m+1})$$

$$= ma_{m+1} - \sum_{k=1}^{m} a_k + \sum_{k=m+2}^{2m+1} a_k - ma_{m+1}$$

$$= \sum_{k=m+2}^{2m+1} a_k - \sum_{k=1}^{m} a_k.$$

Now suppose that $a_l \le x \le a_{l+1}$ where $l \le m$. Then

$$f(x) = \sum_{k=1}^{2m+1} |x - a_k| = \sum_{k=1}^{l} x - a_k = \sum_{k=l+1}^{2m+1} a_k - x$$

$$= \sum_{k=l+1}^{2m+1} a_k - \sum_{k=1}^{l} a_k - (2m - 2l + 1)x.$$

So that

$$f(x) - f(a_{m+1}) = \left[\sum_{k=l+1}^{2m+1} a_k - \sum_{k=1}^{l} a_k - (2m - 2l + 1)x \right]$$

$$- \left[\sum_{k=m+2}^{2m+1} a_k - \sum_{k=1}^{m} a_k \right]$$

$$= 2 \sum_{k=l+1}^{m} (a_k - x) + (a_{m+1} - x) \ge 0.$$

In an analogous way, we see $f(x) \ge f(a_{m+1})$ if $x \le 0$, $x \ge a_n$, or if $a_l \le x \le a_{l+1}$ and $l \ge m + 1$.

If $n = 2m$, and $a_m \le x \le a_{m+1}$, then

$$f(x) = \sum_{k=1}^{2m} |x - a_k| = \sum_{k=1}^{m} x - a_k + \sum_{k=m+1}^{2m} a_k - x = \sum_{k=m+1}^{2m} a_k - \sum_{k-1}^{m} a_k.$$

Now if $a_l \le x \le a_{l+1}$, where $l \ge m + 1$, then

$$f(x) = \sum_{k=l+1}^{2m} a_k - \sum_{k=1}^{l} a_k + (2l - 2n)x$$

and

$$f(x) - \left[\sum_{k=m+1}^{2m} a_k - \sum_{k=1}^{m} a_k \right] = 2 \sum_{k=m+1}^{l} x - a_k \ge 0.$$

Analogously, we can show

$$f(x) \ge \sum_{k=m+1}^{2m} a_k - \sum_{k=1}^{m} a_k,$$

if $x \le a_0$, $x \ge a_n$, or if $a_l \le x \le a_{l+1}$ for $l \le m - 1$.

Look under Real-Valued Functions in the Index for similar problems.

SI978-5

We distinguish three cases. If A, B, C, and D lie on a line, this is problem 1978-4 restated; any point on the middle segment solves the problem.

If A, B, C, and D are the vertices of a convex quadrilateral, suppose AC and BD are the diagonals and that they intersect at O. Let O' be any other point on the plane. Since AOC is a straight line and O is between A and C,

$$|AO| + |OC| = |AC| \leq |AO'| + |O'C|.$$

Similarly,

$$|BO| + |OD| = |BD| \leq |BO'| + |O'D|.$$

Thus,

$$|AO| + |OC| + |BO| + |OD| \leq |AO'| + |O'C| + |BO'| + |O'D|.$$

If A, B, C, and D are not the vertices of a convex quadrilateral, then one of the points, say D, is contained in the triangle whose vertices are the other three points, and $O = D$. Let O' be any point in the plane other than D. For one pair of the points A, B, and C, the triangle formed by this pair and O' contains the triangle formed by this pair and D, so let's say $AO'C$ contains ADC, so that

$$|AO'| + |O'C| \geq |AD| + |DC|.$$

By the triangle inequality, we have

$$|O'B| + |O'D| \geq |BD|,$$

so we then have

$$|AO'| + |O'C| + |BO'| + |O'D| \geq |AD| + |BD| + |CD|.$$

Look under Geometry in the Index for similar problems.

SI978-6

The first step in this solution is to realize that no experiment that always finishes after N tosses of the coin can have probability of success $1/3$. (That is, there are 2^N possible outcomes of flipping a coin N times, and $k/2^N \neq 1/3$ for any integer k. Thus the solution should be an experiment which a patient experimenter would expect to finish in a finite amount of time with probability $1/3$. One such experiment is:

Toss a coin until (a) the coin has been tossed an even number of times and (b) at least one head has been tossed. The experiment is a success if two heads have been tossed.

Thus the experiment consists of two tosses, or four tosses, or six tosses, or .... The experiment is over after two tosses unless both are tails; after four tosses unless all four are tails; and so forth. The experiment will terminate unless the experimenter tosses tails forever, an event whose probability is zero.

The probability that the event is a success is given by

$\Pr(\text{success})$

$$= \sum_{n=1}^{\infty} \Pr(\text{success given that the experiment requires exactly } 2n \text{ tosses})$$

$$\times \Pr(\text{experiment requires exactly } 2n \text{ tosses})$$

$$= \sum_{n=1}^{\infty} \frac{1}{3} \cdot \frac{3}{4^n} = \frac{1}{3}.$$

Look under Probability in the Index for similar problems.

Exam #14–1979

S1979-1

Let M_0 (resp., B_0) be the age of the monkey's mother (resp., the monkey) at the present time; let M_1 (resp., B_1) be the age of the monkey's mother (resp., the monkey) at the time t_1; let M_2 (resp., B_2) be the age of the monkey's mother (resp., the monkey) at the time t_2; and let M_3 (resp., B_3) be the age of the monkey's mother (resp., the monkey) at the time t_3. Let W_w be the weight of the weight in ounces, W_m be the weight of the monkey in ounces, L be the length of the rope, L_m be the length of the rope on the side of the monkey, and L_w be the length of the rope on the side of the weight. Then we have

(1) $M_i - B_i = k$, a fixed difference in ages,

(2) $M_0 + B_0 = 4$,

(3) $M_0 = 2B_1$,

(4) $M_1 = \frac{1}{2}B_1$,

(5) $B_2 = 3M_3$,

(6) $M_3 = 3B_3$.

Since $B_3 = M_3 - k$, equation (6) gives $M_3 = (3/2)k$. So by (5) $B_2 = (9/2)k$ and by (4) $M_1 = (9/4)k$. Therefore

$$B_1 = M_1 - k = (9/4)k - k = (5/4)k$$

and by (3) $M_0 = (5/2)k$. But by (1) and (2), $M_0 + M_0 - k = 4$, so $k = 1$ and $M_0 = 5/2$.

Thus $W_m = (5/2) \cdot 16 = 40$ oz, $W_w = (3/2) \cdot 40 = 60$ oz, and $L = \frac{60}{4} = 15$ ft. Finally, $p = 3/4$ of the rope is on the side of the monkey when the system is at rest since we have

$$60 + 40(1 - p) = 40 + 40p.$$

S1979-2

If $\sin x = \sin y$, then

$$0 = \sin x - \sin y = 2 \cos \frac{x+y}{2} \sin \frac{x-y}{2}.$$

Therefore

$$\frac{x+y}{2} = \frac{\pi}{2} + n\pi \quad \text{or} \quad \frac{x-y}{2} = m\pi,$$

where m and n are integers. The graph thus consists of the two families of lines $y = x + \pi + 2n\pi$ and $y = -x + 2m\pi$, where m and n are integers.

Look under Trigonometry in the Index for similar problems.

S1979-3

Let $V(t)$ be the volume of snow removed at time t, $d(t)$ be the depth of the snow, $a = \frac{dV}{dt}$ be the rate at which snow is removed, $b = \frac{dd}{dt}$ be the rate at which snow falls, and $S(t)$ be the location of the plow. Then $dV = kd(t)ds$, so $\frac{dV}{ds} = kd(t)$. Since

$$\frac{dV}{dt} = \frac{dV}{ds}\frac{ds}{dt}, \quad \frac{ds}{dt} = \frac{a}{kd(t)}.$$

Noting that $d(t) = bt + c$, if t_0 is the time at which it starts snowing, then $t_0 = -c/b$. Suppose $t = 0$ at noon. We have

$$1 = S(1) - S(0) = \int_0^1 \frac{ds}{dt} = \int_0^1 \frac{abdt}{kbd(t)} = k^* \ln\left(d(t)\right)\big|_0^1$$

$$= k^*\left(\ln(c+b) - \ln(c)\right) = k^* \ln\left(1 + \frac{b}{c}\right) = k^* \ln\left(1 - \frac{1}{t_0}\right).$$

(Letting $k^* = a/k$.) Similarly,

$$3/2 = S(2) - S(0) = k^* \ln\left(1 - \frac{2}{t_0}\right).$$

Therefore multiplying corresponding sides of these two equations gives us

$$3k^* \ln\left(1 - \frac{1}{t_0}\right) = 2k^* \left(\ln 1 - \frac{2}{t_0}\right).$$

Solving for t_0, we get $t_0 = \frac{1 \pm \sqrt{5}}{2}$. Since it started snowing before noon, t_0 must be negative and therefore we choose $t_0 = -0.618$ hours. We convert this to minutes by multiplying by 60, to get 37 minutes before noon. Therefore it started snowing at 11:23 a.m.

Look under Differential Equations in the Index for similar problems.

S1979-4

Let r be the radius of the original sphere. Drill the hole along the x-axis, as illustrated in the figure below. Then

$$V = \int_{\sqrt{r^2-9}}^{r} 2\pi y \cdot (2\sqrt{r^2 - y^2})\, dy$$

$$= -2\pi \frac{(r^2 - y^2)^{3/2}}{(3/2)} \Bigg|_{\sqrt{r^2-9}}^{r}$$

$$= \frac{4\pi}{3} \cdot 9^{3/2} = 36\pi.$$

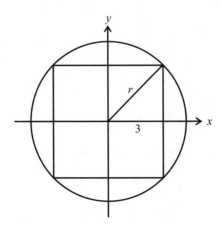

Look under Volumes in the Index for similar problems.

S1979-5

The next several positive integers are:

$$5 = 1 \cdot 9 - 7 + \sqrt{9},$$
$$6 = 1 + 9 - 7 + \sqrt{9},$$
$$7 = 1 \cdot \sqrt{9 + 7} + \sqrt{9},$$
$$8 = 1 - 9 + 7 + 9,$$
$$9 = (-1)^9 + 7 + \sqrt{9},$$
$$10 = 1 + 9 + \lfloor 7/9 \rfloor.$$

This list can be extended to at least 38.

S1979-6

The set of points consists of

$(-6, -1)$, $(-5, -3)$, $(-4, -5)$, $(-4, -4)$, $(-4, 1)$, $(-3, -7)$,
$(-3, -6)$, $(-3, 1)$, $(-3, 2)$, $(-3, 3)$, $(-3, 4)$, $(-3, 5)$, $(-2, -9)$,
$(-2, -8)$, $(-1, -16)$, $(-1, -10)$, $(-1, -2)$, $(0, -13)$, $(0, -2)$,
$(1, -15)$, $(1, -2)$, $(2, -17)$, $(2, -2)$, $(3, -19)$, $(3, -2)$.

Look under Analytic Geometry in the Index for similar problems.

Exam #15–1980

S1980-1

We have

$$a_n b_n - a_{n+1} b_{n+1} \geq \alpha \, a_{n+1}, \quad \text{for } n \geq 1,$$

hence

$$a_1 b_1 > a_1 b_1 - a_{N+1} b_{N+1}$$
$$= \sum_{n=1}^{N} (a_n b_n - a_{n+1} b_{n+1})$$
$$\geq \alpha \sum_{n=1}^{N} a_n + 1.$$

Look under Inequalities in the Index for similar problems.

S1980-2

The problem is to find the weights $w_1 < w_2 < w_3 < w_4$ such that every positive integer $l \leq 40$ can be written in the form $l = \sum_{i=1}^{4} x_i w_i$, where the x_i is either -1, 0, or 1. (A given weight can be placed on the load pan, not used, or on the weight pan to achieve a balance.)

It seems reasonable to choose $w_1 = 1$ and $w_1 + w_2 + w_3 + w_4 = 40$. To measure a 2 pound load, we need $w_2 = 3$ $(l + w_1 = w_2)$. Then we can measure the loads $l = 1, 2, 3, 4$. As $5 + 3 + 1 = 9$, we can also measure $l = 5$ if we take $w_3 = 9$. Then $w_4 = 40 - \sum_1^3 w_i = 27$.

Now note that if

$$0 = \sum_1^4 z_i w_i = z_1 + z_2 w_2 + z_3 w_3 + z_4 w_4,$$

with each $|z_i| \leq 2$, then 3 divides z_1 and hence $z_1 = 0$. So $0 = 3(z_2 + 3z_3 + 3^2 z_4)$ and this gives $z_2 = 0$. Similarly, $z_3 = z_4 = 0$. Therefore, the set

$$S = \left\{ \sum_{i=1}^{4} x_i w_i : -1 \leq x_i \leq 1 \right\}$$

has exactly $3^4 = 81$ members.

As $l \in S$ implies $-l \in S$, it follows that there are exactly 40 positive integers in S and each of these is at most $\sum_1^4 w_i = 40$. So S coincides with the set of integers $1, 2, \ldots, 40$.

Look under Number Theory in the Index for similar problems.

S1980-3

Choose x_0 in $[a, b]$ such that $f(x_0)$ is the maximum value of f on $[a, b]$. Then $f'(x_0) = 0$ and we must also have $f''(x_0) \leq 0$. Therefore

$$0 = f(a) \leq f(x_0) = f'' \leq 0,$$

so $f(x_0) = 0$. That is $f(x) \leq 0$ for all x in $[a, b]$. A similar argument shows that $f(x) \geq 0$ on $[a, b]$ and hence f vanishes identically on $[a, b]$. Notice that this argument assumes that x_0 is in the interior of $[a, b]$, but the same conclusions can be drawn if $x_0 = a$ or b.

Look under Differential Equations in the Index for similar problems.

S1980-4

For

$$x = \begin{pmatrix} x_1 \\ x_2 \\ \vdots \\ x_n \end{pmatrix} \quad \text{and} \quad y^t = (y_1 \quad y_2 \quad \cdots \quad y_n),$$

the matrix

$$A = xy^t = \begin{pmatrix} x_1 y_1 & x_1 y_2 & \cdots & x_1 y_n \\ x_2 y_1 & x_2 y_2 & \cdots & x_2 y_n \\ \vdots & \vdots & \ddots & \vdots \\ x_n y_1 & x_n y_2 & \cdots & x_n y_n \end{pmatrix}.$$

(We can assume $y \neq 0$, as the result is trivially true otherwise.) A short computation shows that if

$$z = \begin{pmatrix} z_1 \\ z_2 \\ \vdots \\ z_n \end{pmatrix},$$

then the vector $Az = (y, z)x$ where $(\cdot, \cdot)$ denotes the usual inner product in $\mathbf{R}^n$. If $z^{(1)}, z^{(2)}, \ldots, z^{(n)}$ is an orthonormal basis with $z^{(i)} = y/\|y\|$, then, with respect to this basis, the matrix $I + A$ has the form

$$I + A = \begin{pmatrix} (z^{(1)}, y)(x, z^{(1)}) + 1 & 0 & 0 & \cdots & 0 \\ (z^{(1)}, y)(x, z^{(2)}) & 1 & 0 & \vdots & 0 \\ \vdots & & 0 & \ddots & 0 \\ (z^{(1)}, y)(x, z^{(n)}) & 0 & \cdots & \cdots & 1 \end{pmatrix}$$

so the determinant of $I + A$ is the product of the diagonal entries, that is

$$\det(I + A) = (z^{(1)}, y)(x, z^{(1)}) + 1$$
$$= (y/\|y\|, y)(x, y/\|y\|) + 1 = (x, y) + 1.$$

Another proof can be given based on the fact that the determinant is unchanged on adding a multiple of one column to another. With this, one reduces A to a lower triangular matrix and then $\det(I + A)$ is easy to compute.

Look under Matrix Algebra in the Index for similar problems.

S1980-5

$$\frac{f(x)}{f(y)} \le 2^{(x-y)^2} \Rightarrow \log_2 f(x) - \log_2 f(y) \le (x-y)^2.$$

If we fix $x = x_0$, divide by $y - x_0$, with $y > x_0$, and then let y tend to x_0, we get

$$\frac{f'(x_0)}{f(x_0)} = \frac{d}{dx} \log_2 f(x)\Big|_{x=x_0} \le 0.$$

A similar argument with $y \le x_0$ gives $f'(x_0)/f(x_0) \ge 0$. Hence $f'(x_0) = 0$ for all x_0, so that f must be a constant function. Conversely, any constant function satisfies the condition.

Look under Real-Valued Functions in the Index for similar problems.

S1980-6

(a) The improper integral has two apparent singularities, at $x = 0$ and $x = \infty$. But as the function

$$f(x) = \begin{cases} \frac{\sin x}{x} & x \ne 0 \\ 1 & x = 0 \end{cases}$$

is continuous on $[0, \infty]$, the integral $\int_0^1 \frac{\sin x}{x}\, dx$ makes sense. So we need only check that

$$\int_1^\infty \frac{\sin x}{x}\, dx = \lim_{b \to \infty} \int_1^b \frac{\sin x}{x}\, dx$$

exists and is finite.

Now

$$\int_1^b \frac{\sin x}{x}\, dx = -\frac{\cos x}{x}\Big|_1^b - \int_1^b \frac{\cos x}{x^2}\, dx$$

$$= \cos 1 - \frac{\cos b}{b} - \int_1^b \frac{\cos x}{x^2}\, dx$$

and $\left|\frac{\cos b}{b}\right| \le 1/b \to 0$. If $g(x) = \cos x/x^2$, then

$$\int_1^b |g(x)|\, dx \le \int_1^b \frac{1}{x^2}\, dx = 1 - \frac{1}{b}$$

so that $\lim_{b \to \infty} \int_1^b |g|$ exists and is finite. This implies that $\lim_{b \to \infty} \int_1^b g(x)\, dx$ exists and is finite by an argument similar to the one used to show that an absolutely convergent series is convergent.

Put $g^+ = \frac{1}{2}(|g| + g)$, $g^- = \frac{1}{2}(|g| - g)$. Then $g^+ \ge 0$, $g^- \ge 0$, so that $\int_1^b g^+$ and $\int_1^b g^-$ are increasing functions of b that are each bounded above by $\int_1^\infty |g|$, hence $\lim_{b\to\infty} \int_1^b g^+$, $\lim_{b\to\infty} \int_1^b g^-$ both exist and are finite so that

$$\int_1^b g = \int_1^b g^+ - \int_1^b g^-$$

has a finite limit as $b \to \infty$. It now follows that $\int_0^\infty \frac{\sin x}{x}\, dx$ is convergent.

(b) Fix $\alpha \geq 0$. Then

$$\int_0^\infty \frac{\sin \alpha x}{x}\, dx = \lim_{b\to\infty} \int_0^b \frac{\sin \alpha x}{x}\, dx = \lim_{b\to\infty} \int_0^{\alpha b} \frac{\sin y}{y}\, dy = \int_0^\infty \frac{\sin x}{x}\, dx.$$

Similarly, if $\alpha \leq 0$,

$$\int_0^\infty \frac{\sin \alpha x}{x}\, dx = -\int_0^\infty \frac{\sin x}{x}\, dx.$$

Therefore, $\left(\int_0^\infty \frac{\sin \alpha x}{x}\, dx \right)^2 = \left(\int_0^\infty \frac{\sin x}{x}\, dx \right)^2$ is constant for $\alpha \neq 0$. Hence its derivative with respect to α is 0 for any $\alpha \neq 0$. The derivative fails to exist at $\alpha = 0$ because of the discontinuity at that point.

Look under Integration in the Index for similar problems.

S1980-7

We have

$$S(2n) = \sum_{k=1}^{2n} \frac{(-1)^{k+1}}{k} = 1 - \frac{1}{2} + \frac{1}{3} - \cdots - \frac{1}{2n}$$

$$= 1 + \left(\frac{1}{2} - \frac{2}{2} \right) + \frac{1}{3} + \left(\frac{1}{4} - \frac{2}{4} \right) + \cdots + \left(\frac{1}{2n} - \frac{2}{2n} \right)$$

$$= \sum_{k=1}^{2n} \frac{1}{k} - \sum_{k=1}^{n} \frac{2}{2k} = \sum_{k=1}^{n} \frac{1}{k} + \sum_{k=1}^{n} \frac{1}{n+k} - \sum_{k=1}^{n} \frac{1}{k}$$

$$= \sum_{k=1}^{n} \frac{1}{n+k}.$$

Look under Finite Sums in the Index for similar problems.

Exam #16–1981

S1981-1

If
$$\frac{3}{8} = \frac{a_0}{7} + \frac{a_1}{7^2} + \frac{a_2}{7^3} + \cdots,$$

then
$$7 \cdot \frac{3}{7} = a_0 + \frac{a_1}{7} + \frac{a_2}{7^2} + \cdots$$

so that a_0 must be the greatest integer in $\frac{21}{8}$, that is, $a_0 = 2$. Consequently,

$$\frac{21}{8} = \frac{16}{8} + \frac{a_1}{7} + \frac{a_2}{7^2} + \cdots, \quad \text{or} \quad \frac{5}{8} = \frac{a_1}{7} + \frac{a_2}{7^2} + \cdots.$$

Since $7 \cdot \frac{5}{8} = a_1 + \frac{a_2}{7} + \cdots$, a_1 must be the greatest integer in $\frac{35}{8}$, so $a_1 = 4$. Continuing in this manner, we find $a_2 = 2$, $a_3 = 4$, $a_4 = 2$. Apparently, $\left(\frac{3}{8}\right)_{10} = (\overline{.24})_7$.

For an alternate solution, we have

$$\left(\frac{3}{8}\right)_{10} = \frac{3_{10}}{8_{10}} = \frac{3_7}{11_7},$$

from which we can obtain the result by long division in base 7.

Look under Number Theory in the Index for similar problems.

S1981-2

Letting A denote the number of acres, G_0 denote the initial amount of grass per acre, S the number of steers, t the number of weeks until the grass is depleted, v the growth rate of the grass per acre per week, and r the rate of consumption of the grass per week, we have the relationship

$$G_0 A + Avt = Srt.$$

The information given yields the following equations:

(1) $G_0 \cdot 10 + 10 \cdot v \cdot 16 = 12 \cdot r \cdot 16$,

(2) $G_0 \cdot 10 + 10 \cdot v \cdot 8 = 18 \cdot r \cdot 8$,

(3) $G_0 \cdot 40 + 40 \cdot v \cdot 6 = S \cdot r \cdot 6$.

Subtracting equation (2) from (1) gives $40v = 24r$, and subtracting equation (3) from four times equation (1) gives $200v = 384r - 3Sr$. Solve for v, substitute, and cancel the r, to find $S = 88$ steers.

Look under Systems of Equations in the Index for similar problems.

S1981-3

Draw a plane passing through the center C of the given sphere, the corner O, and the point P of tangency of the given sphere with one side of the corner. On this plane, we have the trace of the given sphere and the trace of the sphere whose radius we seek, as illustrated below. Since the radius of the given sphere is 1, we know that $\overline{CO} = \sqrt{3}$. If C' is the center of the sphere of radius r that we seek, then $\overline{C'O} = r\sqrt{3}$. Thus

$$\overline{CO} = \overline{CQ} + \overline{QC'} + \overline{C'O},$$

$$\sqrt{3} = 1 + r + r\sqrt{3},$$

$$\sqrt{3} - 1 = r(1 + \sqrt{3}),$$

$$r = \frac{\sqrt{3} - 1}{\sqrt{3} + 1}.$$

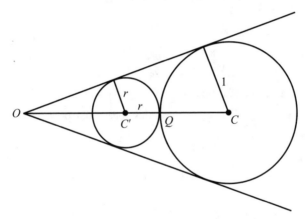

Look under Geometry in the Index for similar problems.

S1981-4

The probability of winning the world series in exactly n games is

$$\frac{1}{2} \cdot \frac{\binom{n-1}{3}}{2^{n-1}} \cdot 2 = \frac{1}{2^{n-1}} \binom{n-1}{3},$$

since $1/2$ is the probability of winning the nth game, $\binom{n-1}{3}/2^{n-1}$ is the probability of winning exactly 3 of the previous $n - 1$ games, and we have 2 teams to choose from. We have $\Pr(n = 4) = \frac{1}{8}$, $\Pr(n = 5) = \frac{1}{4}$, $\Pr(n = 6) = \frac{5}{16}$, $\Pr(n = 7) = \frac{5}{16}$, which sum to one.

Look under Probability in the Index for similar problems.

S1981-5

Let y be the position at time t, measured from the horizontal. Then $y = 50 - 16t^2$, $\frac{dy}{dt} = -32t$, $\theta = \tan^{-1}\frac{y}{100}$, and by the chain rule,

$$\frac{d\theta}{dt} = \frac{d\theta}{dy}\frac{dy}{dt} = \frac{100}{y^2 + 100}(-32t).$$

Therefore

$$\frac{d^2\theta}{dt^2} = \frac{100}{y^2 + 100^2}(-32) - 32t\frac{-100(2y)}{(y^2 + 100^2)^2}\frac{dy}{dt} = 0.$$

Substituting for $\frac{dy}{dt}$, and then y, after some simplification we arrive at $3125 + 400t^2 - 192t^4 = 0$. By the quadratic formula, $y^2 = 125/24$, and so $y = -100/3$ feet below the observer's eye level.

Look under Max/Min Problems in the Index for similar problems.

S1981-6

Suppose the path of the destroyer is given by

$$x = r\cos\theta,$$

$$y = r\sin\theta,$$

where $r = r(\theta)$ is some function of θ. If s denotes arclength along the destroyer's path as illustrated below, then

$$(ds)^2 = (dx)^2 + (dy)^2.$$

Since $dx = -r\sin\theta\,d\theta + \cos\theta\,dr$, and $dy = r\cos\theta\,d\theta + \sin\theta\,dr$,

$$(ds)^2 = r^2(d\theta)^2 + (dr)^2.$$

Since the submarine is moving at a constant rate c, the distance r that it travels in a straight line is given by $r = ct$. Equating the distance $s + 2$ that the destroyer has gone at the time t of interception with the distance $2ct$ that the submarine has gone, we find $s + 2 = 2ct$. Thus $s + 2 = 2r$, so $ds = 2dr$. Consequently,

$$(ds)^2 = 4(dr)^2 = r^2(d\theta)^2 + (dr)^2$$

and thus $3(dr)^2 = r^2(d\theta)^2$. Taking square roots, separating the variables, and integrating gives $r = K\exp(\theta/\sqrt{3})$ for some constant K. Since $r = 1$ when $\theta = 0$, $K = 1$.

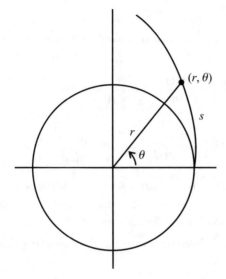

Look under Analytic Geometry in the Index for similar problems.

Exam #17–1982

S1982-1

Let x_1, x_2, x_3 be the roots of $x^3 + ax^2 + bx + c$. It is known that the coefficients of a polynomial are symmetric functions in the roots. Let

$$s_1 = x_1 + x_2 + x_3,$$

$$s_2 = x_1 x_2 + x_1 x_3 + x_2 x_3,$$

$$s_3 = x_1 x_2 x_3.$$

Then we have

$$c = -s_3 = -x_1 x_2 x_3,$$

$$b = s_2 = x_1 x_2 + x_1 x_3 + x_2 x_3,$$

$$a = -s_1 = -(x_1 + x_2 + x_3).$$

Let a', b', c' be the corresponding coefficients of the cubic whose roots are reciprocals of the original cubic. Then

$$c' = -\frac{1}{x_1}\frac{1}{x_2}\frac{1}{x_3} = \frac{1}{c},$$

$$b' = \frac{1}{x_1 x_2} + \frac{1}{x_1 x_3} + \frac{1}{x_2 x_3} = \frac{x_1 + x_2 + x_3}{x_1 x_2 x_3} = \frac{a}{c},$$

and

$$a' = -\left(\frac{1}{x_1} + \frac{1}{x_2} + \frac{1}{x_3}\right) = -\frac{x_1 x_2 + x_1 x_3 + x_2 x_3}{x_1 x_2 x_3} = \frac{b}{c}.$$

Therefore the desired cubic is

$$x^3 + \frac{b}{c}x^2 + \frac{a}{c}x + \frac{1}{c}.$$

Look under Polynomials in the Index for similar problems.

S1982-2

A separated subset of order k from a set of order n can be represented by a sequence of k dashes and $n - k$ slashes where no two dashes are adjacent. For example, the separated subset $\{2, 5, 7\}$ with $n = 8$ can be represented as /-//-/-/. Given a set of $n - k$ slashes, and conditions of this problem, there are $n - k + 1$ places to possibly place the k dashes (this includes before the first slash and after the last slash). Thus there are $\binom{n-k+1}{k}$ different separated subsets.

Look under Enumeration in the Index for similar problems.

S1982-3

Let the two points that are furthest apart be denoted A and B. Join them by a line, wlog horizontal. Now draw the smallest rectangle containing all of the points with top and bottom sides parallel to the line AB, vertical sides through A and B. There will be points C and D on the top and bottom, respectively, of the rectangle. We consider the quadrilateral formed by $ACBD$. The area of triangle ACB is $\frac{1}{2}h\overline{AB} \le 1$ and the area of triangle ABD is $\frac{1}{2}h'\overline{AB} \le 1$, where h and h' are the altitudes of the triangles. Thus the area of the rectangle is

$$(h + h')\overline{AB} = h\overline{AB} + h'\overline{AB} \le 2 + 2 = 4.$$

Look under Geometry in the Index for similar problems.

S1982-4

$f(x)$ is differentiable at $x = 0$ if

$$\lim_{h \to 0} \frac{f(h) - f(0)}{h} = \lim_{h \to 0} \frac{e^{-1/h^2}}{h}$$

exists. Set $y = 1/h^2$, then

$$\lim_{h \to 0} \frac{e^{-1/h^2}}{h} = \lim_{y \to \infty} y^{1/2} e^{-y} = 0.$$

Look under Differentiation in the Index for similar problems.

S1982-5

If $\alpha \leq 1$, then f is uniformly continous on its domain.

Look under Real-Valued Functions in the Index for similar problems.

S1982-6

If a^3 ends in the digits 11, then $a^3 \equiv 11 \pmod{100}$. This is equivalent to the two conditions $a^3 \equiv 11 \equiv 3 \pmod 4$ and $a^3 \equiv 11 \pmod{25}$. By an easy calculation, 3 is the only residue satisfying $a^3 \equiv 11$ modulo 4. Now, $a^3 \equiv 11 \pmod{25} \Rightarrow a^3 \equiv 1 \pmod 5$. By another easy calculation, $a = 1$ is the only residue modulo 5 which satisfies this equation. That means we need only check the cubes of 1, 6, 11, 16, and 21 modulo 25. $1^3 = 1, 6^3 = 36 \cdot 6 = 11 \cdot 6 = 16, 16^3 = 256 \cdot 16 = 6 \cdot 16 = 96 = 21,$ $21^3 = (-4)^3 = -64 = -14 = 11$. So $a = 21$ is the only solution modulo 25. By the Chinese remainder theorem, there is only one number satisfying $a \equiv 3 \pmod 4$ and $a \equiv 21 \pmod{25}$ (that number is 71). It follows that the probability that a random cube ends with the digits 11 is $1/100$.

Look under Number Theory or Probability in the Index for similar problems.

S1982-7

Using the disk method for finding volumes of revolution, we obtain the integral

$$\pi \int_{-a}^{a} \left(\left(\sqrt{a^2 - y^2} + b \right)^2 - b^2 \right) dy = \pi^2 \cdot a^2 \cdot b + \frac{4\pi a^3}{3}.$$

Look under Volumes in the Index for similar problems.

Exam #18–1983

S1983-1

$$\lim_{n\to\infty} \sqrt[n]{n!} > \lim_{n\to\infty} \sqrt[n]{(n/2)^{n/2}} = \lim_{n\to\infty} \sqrt{n/2} = \infty$$

Look under Limit Evaluation in the Index for similar problems.

S1983-2

There are 2^{10} possible outcomes when one flips a coin 10 times. Let T_i be the result of the ith flip. We want to compute the number of possible outcomes which have $T_9 = T_{10} = H$ and no adjacent H's in $T_1, T_2, \ldots, T_8, T_9$. Certainly then, $T_8 = T$ and we wish to know how many sequences of length 7 have no adjacent H's. Let S_n be the number of sequences of length n without 2 consecutive occurrences of H. Each such sequence must either start with HT followed by a sequence of length $n - 2$ with no consecutive H's or start with T followed by a sequence of length $n - 1$ with no consecutive H's. Therefore $S_n = S_{n-1} + S_{n-2}$. Also, $S_1 = 2, S_2 = 3$. We see then that $S_n = F_{n+2}$ where F_i denotes the ith Fibonacci number. Hence $S_7 = F_9 = 34$. It follows that there are 34 possible outcomes which satisfy our condition. The probability that such an outcome occurs is then $34/1024$. (We have solved a more general version of the problem: if 10 is replaced by an arbitrary $n \in \mathbf{N}$, then the probability that we stop after the nth flip is $F_{n-1}/2^n$.)

Look under Fibonnacci Sequences or Probability in the Index for similar problems.

S1983-3

From the illustration below, we want to maximize

$$f(x) = (a - x)x + \pi r^2.$$

Since we have the relationship

$$\frac{1}{2}(a - x)(a - x) = \frac{1}{2}r\sqrt{2}(a - x) + \frac{1}{2}r(a - x) + \frac{1}{2}r(a - x),$$

we see that

$$r = \frac{(a - x)(a - x)}{(a - x) + (a - x) + \sqrt{2}(a - x)} = \frac{a - x}{2 + \sqrt{2}}.$$

Thus we want to maximize

$$f(x) = (a - x)x + \frac{\pi(a - x)^2}{(2 + \sqrt{2})^2},$$

where $0 \leq x \leq 1$. Letting $A = \pi/(2 + \sqrt{2})^2$, we can write

$$f(x) = x^2(A - 1) + x(a - 2aA) + Aa^2.$$

Differentiating gives us

$$f'(x) = 2x(A - 1) + a(1 - 2A),$$

which has a critical point at $x = a(1 - 2A)/2(1 - A)$. Further this is a maximum since $A \leq 1$ gives us $f''(x) = 2(A - 1) \leq 0$. The dimensions of the rectangle and circle easily follow.

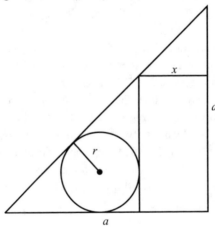

Look under Geometry or Max/Min Problems in the Index for similar problems.

S1983-4

Solution 1. Suppose that $t \in \mathbf{R}$ is not a finite sum of elements from $S = \{1, 1/2, 1/3, 1/4, \cdots\}$. Define $f(x) = \max\{\frac{1}{n} \in S : \frac{1}{n} \leq x\}$. Since zero is a limit point of S, f is well defined for $x > 0$. Finally, define $t_0 = f(t), t_{i+1} = t_i + f(t - t_i)$. Since t is not a finite sum of elements from S, we must have $t - t_i \neq 0$ for all i. Now, since f only attains positive values, the sequence $\{t_i\}$ is monotone increasing. Furthermore, since $f(t - t_i) < t - t_i$, we have $t_i < t$ for all i so that $\{t_i\}$ is bounded above. Every bounded monotone sequence converges, so $t_i \to s$ for some

$s \in \mathbf{R}$. If $s < t$, then let $\frac{1}{m} = f(t - s)$. $t_i \to s$, so there exists a t_i for which $s - t_i < \frac{1}{m}$. It follows immediately that $t_{i+1} \geq t_i + \frac{1}{m} > s$, a contradiction. Therefore $t = s$.

Solution 2. Case 1: assume that our positive number, p, satisfies $0 < p < 1$. Then p has a binary representation of the form

$$p = 0.b_1 b_2 b_3 \ldots = b_1 \frac{1}{2} + b_2 \frac{1}{4} + b_3 \frac{1}{8} + \cdots,$$

with each b_k either zero or one. This gives a representation of p as a sum (possibly infinite) of a subset of the numbers $\{1/2, 1/4, 1/8, \ldots\}$. Case 2: assume that our positive number M satisfies $M \geq 1$. Since the series $\frac{1}{3} + \frac{1}{5} + \frac{1}{7} + \cdots$ diverges, there is a finite sum $S = \sum_{k=1}^{n} \frac{1}{2k+1}$ such that $M = S + q$ with $0 < q < 1$. Applying case 1 to this q gives the result.

Look under Infinite Series in the Index for similar problems.

S1983-5

This is a repeat of problem # 3 on Exam # 7 (1972).

S1983-6:

Assume that there is a solution with $z < n$. Without loss of generality we have $x \leq y$, and therefore $x \leq y < z < n$. From this,

$$x^n = z^n - y^n$$
$$= (z - y)[z^{n-1} + yz^{n-2} + \cdots + y^{n-1}]$$
$$> (1)[x^{n-1} + x^{n-1} + \cdots + x^{n-1}]$$
$$= nx^{n-1}.$$

But now $x^n > nx^{n-1}$ implies $x > n$, a contradiction.

Look under Diophantine Equations in the Index for similar problems.

S1983-7

N is a perfect square.

$$N = (n - 1)n(n + 1)(n + 2) + 1 = n^4 + 2n^3 - n^2 - 2n + 1$$
$$= (n^2 + n - 1)^2.$$

Look under Number Theory in the Index for similar problems.

SI983-8

First, note that $p_k(x_k) = 1$ (this is obvious from the definition) and $p_k(x_j) = 0$ for $j \neq k$ (because of the $x - x_j$ term in the numerator). Let

$$P(x) = \sum_{k=1}^{n} p_k(x) - 1,$$

then $P(x)$ is a polynomial of degree at most $n - 1$ because each $p_k(x)$ is a polynomial of degree $n - 1$. But notice that for each x_j,

$$P(x_j) = p_1(x_j) + \cdots + p_n(x_j) - 1 = p_j(x_j) - 1 = 1 - 1 = 0.$$

Since the x_j are distinct, it follows that $P(x)$ has n distinct zeros. Any polynomial of degree at most $n - 1$ with n zeros must be identically zero, therefore $P(x) = 0$ which implies $\sum_{k=1}^{n} p_k(x) = 1$.

Look under Polynomials in the Index for similar problems.

SI983-9

$g(x) = g(y) \Rightarrow g^m(x) = g^m(y) \Rightarrow x = y$, therefore g is injective. Any injective real-valued continuous function must be strictly monotone on its domain. Suppose that g is increasing on $[0, 1]$, then for $x \in [0, 1]$,

$$x > g(x) \to g(x) > g^2(x) \Rightarrow \cdots \Rightarrow g^{m-1}(x) > g^m(x),$$

hence $x > g^m(x) = x$ which is obviously false. Also,

$$x < g(x) \Rightarrow g(x) < g^2(x) \Rightarrow \cdots \to g^{m-1}(x) < g^m(x),$$

so that $x < g^m(x) = x$. So we must have $g(x) = x$. Now if g is decreasing on $[0, 1]$, then for $x, y \in [0, 1]$, $x < y \Rightarrow g(x) > g(y) \Rightarrow g^2(x) < g^2(y)$ so that g^2 is increasing on $[0, 1]$. Using the above argument we see that $g^2(x) = x$ must hold.

Look under Real-Valued Functions in the Index for similar problems.

Exam #19-1984

SI984-I

By the binomial theorem $1,005,010,010,005,001$ is equal to $(1001)^5 = 7^5(11^5)13^5$.

Look under Number Theory in the Index for similar problems.

S1984-2

Every path from A to B goes along 9 edges, 5 slanting right, 4 slanting left and any such path joins A to B. Thus there are $\binom{9}{5} = 126$ paths.

Look under Enumeration in the Index for similar problems.

S1984-3

Door D is changed by the nth person if and only if n divides D. Doors that have an even number of divisors are closed, those with an odd number are open. If n divides D, then so does D/n, thus D has an even number of divisors unless D is a perfect square. Thus, doors $1, 4, 9, 16, \ldots$ are open, the rest are closed.

Look under Number Theory in the Index for similar problems.

S1984-4

This problem concerns the numerical semigroup

$$S = \{ap + bq \,|\, a, b \text{ are nonnegative integers}\}$$

generated by $p = 7$ and $q = 12$ and asks "What is the largest integer not in S?" (In the general solution here, all variables will be integers and p and q will denote relatively prime positive integers.) In this question the emphasis is on the nonnegativity because, since p and q are relatively prime, every integer can be expressed in the form $ap + bq$ for some integers a and b.

Two fairly easy observations, which we offer without proof, enable one to do the stated problem with a little trial and error.

Lemma 1.   *If n is a positive integer such that for each $k = 0, 1, 2, \ldots,$ $p - 1$ there are nonnegative a_k and b_k with $n + k = a_k p + b_k q$, then every integer greater than n is in the semigroup S.*

Lemma 2.   *Let $m = ap + bq$ where a and b are integers. Then $m = a'p + b'q$ if and only if there is an integer k so that $a' = a + kq$ and $b' = b - kp$.*

Corollary.   *If m is an integer, then there are unique integers a and b with $0 \le a \le q - 1$ so that $m = ap + bq$.*

Now since $65 = (11)7 + (-1)12$ and $65 = 7a + 12b$ with a and b nonnegative implies $a \le 9$, the uniqueness in the corollary shows that it is not possible to give exact change of 65 quanta. Moreover, since $66 = 6(7) + 2(12)$, $67 = 1(7) + 5(12)$, $68 = 8(7) + 1(12)$, $69 = 3(7) + 4(12)$,

$70 = 10(7) + 0(12)$, $71 = 5(7) + 3(12)$, and $72 = 0(7) + 6(12)$, Lemma 1 shows 65 is the largest such price. This is actually a special case of an old (1884) theorem by Sylvester.

Theorem. *If p and q are relatively prime positive integers, then $(p-1)(q-1) - 1$ cannot be written in the form $ap + bq$ for nonnegative integers a and b, but every larger integer can be.* (Proof omitted.)

The following theorem, also by Sylvester in 1884, shows that there are 33 prices for which Subsylvanians cannot give exact change.

Theorem. *If p and q are relatively prime positive integers and S is the numerical semigroup generated by p and q, then there are $(p-1)(q-1)/2$ positive integers not in S.* (Proof omitted.)

Look under Number Theory in the Index for similar problems.

S1984-5

In order for three segments to form the sides of a triangle, it is necessary and sufficient that the sum of the lengths of any two segments must be greater than the length of the third. In our situation, this means that none of the segments formed may have length greater than or equal to $1/2$.

The easiest way to find the required probability is to represent the two points chosen in the interval as an ordered pair from the unit square and decide which pairs lead to triangles. In the figure below, the region corresponding to possible triangles has been shaded.

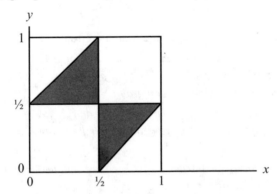

Because of the independence and uniformity assumptions, the probability that the first point is chosen from the interval $[a, b]$ and the second from the interval $[c, d]$ is just $(b-a)(d-c)$. More generally, the probability that the points x and y are chosen from $[0, 1]$, so that (x, y) is in the subset A

of the unit square, is equal to the area of A. Thus the probability that a triangle can be formed is the area of the shaded region in the figure, $1/4$.

We may obtain the same result more analytically with the principles of conditional probability. Recall that if $x_1, x_2, \ldots, x_n$ form a partition of the space of outcomes and z is any event, then

$$\Pr(z) = \sum_{i=1}^{n} \Pr(z|x_i) \Pr(x_i).$$

In our situation, letting X be the random variable for the choice of the first point, then this becomes

$$\Pr(\text{a triangle}) = \int_{x=0}^{1} \Pr(\text{a triangle } X = x) \, dx.$$

Now, when $0 \le x \le 1/2$, in order for a triangle to be formed, the other point must be in the interval $(1/2, 1/2 + x)$, which has probability $(x + 1/2) - x = x$. When $1/2 < x \le 1$, in order for a triangle to be formed, the second point must be in the interval $(x - 1/2, 1/2)$, which has probability $1/2 - (x - 1/2) = 1 - x$. Thus, the probability of having a triangle is

$$\int_{x=0}^{1/2} x \, dx + \int_{x=1/2}^{1} (1 - x) \, dx = 1/4.$$

(This might be interesting to verify with a computer simulation.)

Look under Probability in the Index for similar problems.

S1984-6

(a) From the functional equation and the continuity of f at 1, we have

$$\lim_{x \to \infty} f(x) = \lim_{x \to \infty} f\left(x/[x + 1]\right) - 2 = f(1) - 2 = 3.$$

(b) We note first of all that it is not sufficient to observe that $f(1/n) = 2n + 3$ for $n = 1, 2, 3, \ldots$; the function $g(x) = (2/x)\cos(2\pi/x) + 3$ has the property that $g(1/n) = 2n + 3$ for $n = 1, 2, 3, \ldots$ but certainly, $\lim_{x \to 0+} g(x) \neq +\infty$.

Since $f(x)$ is continuous on $0 < x < \infty$, it has a finite minimum value on every bounded closed interval. Let c be the minimum value of f on the interval $[1/(n + 1), 1/n]$. Since $1/(n + 1)$ equals $(1/n)/[(1/n) + 1]$, the functional equation shows that $c_{n+1} = c_{n+2}$, hence that $c_{n+1} = c_1 + 2n$.

Thus, if $0 < x \leq 1/(1+n)$, we have

$$f(x) \geq \inf_{m \geq n+1} c_n = c_{n+1} = c_1 + 2n$$

which means $\lim_{x \to 0^+} f(x) = +\infty$.

(c) The given conditions relate the values of f on the intervals $[1, \infty)$, $[1/2, 1)$, $[1/3, 1/2), \ldots$ to each other and specify the values at the end points, but besides $f(1) = 5$ and $\lim_{x \to \infty} f(x) = 3$, the function f is an arbitrary continuous function on $[1, \infty)$. A succinct way of say this is: if g is any continuous function on $(-\infty, \infty)$ such that $\lim_{x \to \pm\infty} g(x) = 0$, then $f(x) = 3 + (2/x) + g(\cot(\pi/x))$ satisfies the given conditions and any solution can be written in this way for some g.

Look under Limit Evaluation in the Index for similar problems.

S1984-7

It is helpful to consider 25 tons as one unit of weight; we want to count the number of trains of total weight 40 units. Let $f(n)$ denote the number of trains of total weight n units: we want to find $f(40)$.

It is easy to find $f(n)$ when n is small: $f(1) = 1$ (the only train is B); $f(2) = 3$ (BB, F, H); $f(3) = 5$ (BF, BH,BBB,FB,HB); $f(4) = 11$ (BFF, FF, HF, BBH, FH, HH, BFB, BHB,BBB,FBB,HBB); etc. More importantly, we notice that an n unit train is either an $n-1$ unit train with a boxcar stuck on the end or an $n-2$ unit train with a flat car or hopper car stuck on the end. That is, f satisfies $f(n) = f(n-1)+2f(n-2)$ where $f(1) = 1$ and $f(2) = 3$. Thus $f(5) = 11 + 10 = 21$, $f(6) = 21 + 22 = 43$, $f(7) = 43 + 42 = 85$, and so on until we arrive at

$$f(39) = \frac{2^{40} - 1}{3} \quad \text{and} \quad f(40) = \frac{2^{41}}{3},$$

our answer.

(It is easy to prove by induction, but perhaps hard to discover, that $f(n) = f(n-1) + 2f(n-2)$ where $f(1) = 1$ and $f(2) = 3$ implies:

$$f(2k-1) = \frac{2^{2k} - 1}{3} \quad \text{and} \quad f(2k) = \frac{2^{2k+1} + 1}{3}$$

for $k = 1, 2, 3, \ldots$)

Look under Enumeration in the Index for similar problems.

Exam #20–1985

S1985-1

We have

$$x^4 + bx^3 + cx^2 + dx + e = (x - \alpha)(x - \beta)(x - \gamma)(x - \delta).$$

In the above equality, first put $x = i$, and then $x = -i$, where $i = \sqrt{-1}$, to get

$$1 - c + e - (b - d)i = (i - \alpha)(i - \beta)(i - \gamma)(i - \delta),$$

and

$$1 - c + e + (b - d)i = (-i - \alpha)(-i - \beta)(-i - \gamma)(-i - \delta).$$

Noting that $(u + vi)(u - vi) = u^2 + v^2$, multiplication of the corresponding sides of the two equalities above yields

$$(\alpha^2 + 1)(\beta^2 + 1)(\gamma^2 + 1)(\delta^2 + 1) = (1 - c + e)^2 + (b - d)^2.$$

Look under Polynomials in the Index for similar problems.

S1985-2

Since $3333 \equiv 1 \pmod 7$ and $4444 \equiv -1 \pmod 7$, we have

$$3333^{4444} + 4444^{3333} \equiv 1^{4444} + (-1)^{3333} \pmod 7 \equiv 1 - 1$$

$$\equiv 0 \pmod 7.$$

Some more problems of this type:

(i) $2222^{1111} + 1111^{2222}$ is divisible by 3.

(ii) $1111^{4444} + 4444^{1111}$ is divisible by 5.

(iii) $2222^{5555} + 5555^{2222}$ is divisible by 7.

(iv) $88888^{33333} + 33333^{88888}$ is divisible by 11.

Look under Number Theory in the Index for similar problems.

S1985-3

Clearly U is the multiplicative group of units modulo 21. Since 6 is a square-free integer and U is abelian, each subgroup to be found is cyclic. Hence the subgroup along the solid path could be $\{1, 2, 4, 8, 16, 11\}$; the one along the dashed path could be $\{1, 5, 4, 20, 16, 17\}$; and the one along the dotted path could be $\{1, 13, 4, 10, 16, 19\}$.

Look under Group Theory in the Index for similar problems.

S1985-4

There are 2 types of triangles that can be formed: either each vertex lies on a different side of the square, or 2 vertices lie on one side and the third on another side.

We find the number of triangles of the first type as follows: Select 3 of the 4 sides of the square (this can be done in $\binom{4}{3}$ or 4 ways) and then select 1 of the 100 points of subdivision on each of these 3 sides (this can be done in 100^3 ways). Thus, there are $4 \cdot 100^3$ triangles of the first kind.

The number of triangles of the second type is found as follows: Select 1 side of the square (4 ways of doing this) and 2 of the division points on this side ($\binom{100}{2}$ ways) and then select 1 of the 3 remaining sides (3 ways) and 1 of the 100 points of subdivision (100 ways) on this side. The total number of triangles of the type is $4 \cdot 3 \cdot 100 \cdot \binom{100}{2} = 6 \cdot 99 \cdot 100^2$.

Hence the required number is $4 \cdot 100^3 + 6 \cdot 99 \cdot 100^2 = 9{,}940{,}000$.

Look under Enumeration in the Index for similar problems.

S1985-5

From $a_n = (n-1)(a_{n-1} + a_{n-2})$, we get

$$a_n - na_{n-1} = -\big[a_{n-1} - (n-1)a_{n-2}\big].$$

Iterating this relation $n - 3$ times, we have

$$a_n - na_{n-1} = (-1)^{n-2}(a_2 - 2a_1) = (-1)^{n-2} = (-1)^n.$$

Hence,

$$\frac{a_n}{n!} - \frac{a_{n-1}}{(n-1)!} = \frac{(-1)^n}{n!}$$

and

$$\sum_{r=2}^{n}\left(\frac{a_r}{r!} - \frac{a_{r-1}}{(r-1)!}\right) = \sum_{r=2}^{n}\frac{(-1)^r}{r!} = \sum_{r=0}^{n}\frac{(-1)^r}{r!}.$$

Since $a_1 = 0$, the sum on the left telescopes to $\frac{a_n}{n!}$. Thus

$$a_n = n!\left(\sum_{r=0}^{n}\frac{(-1)^r}{r!}\right).$$

and

$$\lim_{n\to\infty} \frac{a_n}{n!} = \lim_{n\to\infty} \sum_{r=0}^{n} \frac{(-1)^r}{r!} = e^{-1}.$$

(Observe that a_n is the number of derangements of a set of n objects .)

Look under Limit Evaluation or Sequences in the Index for similar problems.

S1985-6

We will prove this result by induction. The result is clearly true for $n = 1$. Assume the result true for $n = m$. Let S be a set of $m+2$ integers between 1 and $2m + 2$.

Suppose S does not contain both $2m + 1$ and $2m + 2$. Let S_1 be the subset of S not containing $2m + 1$ or $2m + 2$. Then S_1 contains $m + 1$ integers between 1 and $2m$. By the induction hypothesis, the result is true for S_1 and therefore for S.(Of course if neither $2m + 1$ nor $2m + 2$ is in S, then by the induction assumption, the result is already true for S).

Suppose S contains both $2m + 1$ and $2m + 2$. If $m + 1$ is in S, then $2m + 2$ is a multiple of S, and thus the result is true for S. Therefore, assume $m + 1$ is not in S. Let S_1 be the set obtained from S by throwing out $2m+1$ and $2m+2$ and putting in $m+1$. Then S_1 has $m+1$ integers between 1 and $2m$. By the induction hypothesis, some element b of S_1 is a multiple of some other element a of S_1. If a and b are both in S, then the result is true for S. Now a cannot be $m+1$. Because the only positive integers $m + 1$ can divide other than itself are $k(m + 1)$, where k is an integer ≥ 2. But S_1 does not contain such integers. Hence a must be in S. If b is $m + 1$, then since $2m + 2$ is a multiple of b and b is a multiple of a, $2m + 2$ is a multiple of a. Thus the result is true for S.

Look under Number Theory in the Index for similar problems.

S1985-7

From $x+y = 2a$, we have $y = 2a-x$. Substituting in $xy > [(b^2-1)/b^2]a^2$, we get $b^2x(2a - x) > (b^2 - 1)a^2$,

i.e., $b^2x^2 - 2ab^2x + (b^2 - 1)a^2 < 0$,

i.e., $\big(bx - (b - 1)a\big)\big(bx - (b + 1)a\big) < 0$,

i.e., $\dfrac{(b-1)a}{b} < x < \dfrac{(b+1)a}{a}.$

Thus, the number of favorable cases is

$$\frac{(b+1)a}{b} - \frac{(b-1)a}{b} = \frac{2a}{b}.$$

The number of all possible cases is when $0 < x < 2a$, and hence is $2a$. The required probability is therefore

$$\frac{2a/b}{2a} = \frac{1}{b}.$$

Look under Probability in the Index for similar problems.

Exam #21–1986

S1986-1

See Solution #1 on Exam #14.

S1986-2

Let $B = \{b_1, b_2, b_3\}$. The matrix of ϕ with respect to B is given by

$$A = \begin{pmatrix} 0 & 2i & 0 \\ 2 & 0 & 0 \\ 0 & 0 & 1+i \end{pmatrix}.$$

To find the eigenvalues we set

$$\det(A - \beta I) = 0 = \left[(1+i) - \beta\right]\left[\beta^2 - 4i\right]$$

which implies that $\beta = 1 + i$, $\beta = \sqrt{2} + \sqrt{2}i$ or $\beta = -\sqrt{2} - \sqrt{2}i$.

Look under Complex Numbers or Matrix Algebra in the Index for similar problems.

S1986-3

$i^i = \left(e^{\pi i/2}\right)^i = e^{-\pi/2}$.

Look under Complex Numbers in the Index for similar problems.

S1986-4

For every four distinct points of the N points there is one and only one pair of segments that intersect and have the given four points as endpoints.

Thus the number of intersections is given by:

$$\binom{n}{4} = \frac{n(n-1)(n-2)(n-3)}{24}.$$

Look under Enumeration in the Index for similar problems.

S1986-5

Let $x = \sqrt{2} + (\sqrt[3]{3})$. Then

$$x^2 = 2 + 2\sqrt{2}(\sqrt[3]{3}) + (\sqrt[3]{3})^2,$$

$$x^3 = 2\sqrt{2} + 3(2(\sqrt[3]{3})) + 3\sqrt{2}(\sqrt[3]{3})^2 + 3.$$

Similarly

$$(\sqrt[3]{3})^2 = \sqrt[3]{9} = x^2 - 2 - 2\sqrt{2}\sqrt[3]{3},$$

But $\sqrt[3]{3} = x - \sqrt{2}$, so

$$\sqrt[3]{9} = x^2 - 2 - 2\sqrt{2}(x - \sqrt{2}) = x^2 - 2\sqrt{2}x + 2.$$

Now

$$x^3 = 2\sqrt{2} + 6(x - \sqrt{2}) + 3\sqrt{2}(x^2 - 2\sqrt{2}x + 2) + 3$$

$$= 2\sqrt{2} + 6x - 6\sqrt{2} + 3\sqrt{2}x^2 - 12x + 6\sqrt{2} + 3$$

$$= 2\sqrt{2} + 3\sqrt{2}x^2 - 6x + 3.$$

Thus $x^3 + 6x - 3 = \sqrt{2}(3x^2 + 2)$. Squaring both sides gives

$$x^6 + 36x^2 + 9 + 12x^4 - 6x^-36x = 18x^4 + 24x^2 + 8$$

or

$$x^6 - 6x^4 - 6x^3 + 12x^2 - 36x + 1 = 0.$$

Look under Polynomials in the Index for similar problems.

S1986-6

(a) If the limit of $f(x_1, x_2)$ exists as (x_1, x_2) tends to $(0,0)$, then the limit exists along any path to $(0,0)$. In particular the limit must exist along the path defined by $x_1 = x_2^2$. Then the limit of $f(x_1, x_2)$ as (x_1, x_2) tends to $(0,0)$ now becomes

$$\lim_{x_2 \to 0} f(x_2^2, x_2) = \lim_{x_2 \to 0} \frac{x_2^4}{2x_2^4} = \frac{1}{2} \neq 0.$$

Thus $f(x_1, x_2)$ is not continuous at $(0,0)$.

(b) Since $f(x_1, x_2)$ is not continuous at $(0,0)$, $f(x_1, x_2)$ is not differentiable at $(0,0)$.

Look under Multivariate Calculus in the Index for similar problems.

Exam #22–1987

S1987-1

The well-known proof that there are arbitrarily long sequences of composite integers suggests that the solution should be $11!+2$, $11!+3,\ldots,11!+11$, but a calculation shows that these numbers are much too large. An appropriate idea is present though; we should replace $11!$ by a number divisible by $2, 3, \ldots, 11$. For example, the least common multiple of these integers is $2^3 \cdot 3^2 \cdot 5 \cdot 7 \cdot 11 = 27720$, so the sequence $27722, 27723, \ldots, 27731$ is a solution. There are smaller solutions, the smallest solution is the sequence beginning with 114.

Look under Number Theory in the Index for similar problems.

S1987-2

Each layer is a triangle of cannonballs, the kth having k balls on each edge. Thus, each layer has

$$1 + 2 + \cdots + k = \frac{k(k+1)}{2} = \frac{1}{2}(k^2 + k)$$

cannonballs. The total number of cannonballs in a stack is therefore

$$\frac{1}{2}\left((1^2 + 1) + (2^2 + 2) + \cdots + (n^2 + n)\right)$$

$$= \frac{1}{2}\left((1^2 + 2^2 + \cdots + n^2) + (1 + 2 + \cdots + n)\right)$$

$$= \frac{1}{2}\left(\frac{n(n+1)(2n+1)}{6} + \frac{n(n+1)}{2}\right)$$

$$= \frac{1}{6}n(n+1)(n+2).$$

Look under Enumeration in the Index for similar problems.

S1987-3

If the ball obeys the law exactly, it must bounce an infinite number of times, but it is not clear whether it bounces for an infinite time. We should find the infinite series that represents the total time, decide if it converges, and if it converges compute the sum.

Since it is dropped from a height of 16 feet, it will require $\sqrt{16}/4 = 1$ second to fall to the ground. It then bounces to a height of 8 feet, requiring $\sqrt{8}/4$ second to rise and $\sqrt{8}/4$ second to fall to the ground again. Continuing, we see that the total time is the sum of the series

$$1 + 2(\sqrt{8}/4) + 2(\sqrt{4}/4) + 2(\sqrt{2}/4) + \cdots,$$

or

$$1 + \left(\frac{2}{4}\right) \sum_{j=0}^{\infty} \sqrt{8/2^j} = 1 + \sqrt{2} \sum_{j=0}^{\infty} \left(\frac{1}{\sqrt{2}}\right)^j.$$

Since this is a geometric series with ratio less than 1, the series converges and its sum is

$$1 + \sqrt{2}\frac{1}{1 - 1/\sqrt{2}} = \frac{\sqrt{2}+1}{\sqrt{2}-1}.$$

Thus, this model predicts the ball will bounce slightly more than 5.8 seconds. (In fact, this model is in fair agreement with experiment; the constant $1/2$ in this problem is called the coefficient of restitution of the ball.)

Look under Infinite Series in the Index for similar problems.

S1987-4:

Let S_j denote the event that the jth senior draws his or her own name from the hat. We wish to find the probability that none of these events occurs, that is, we want

$$\Pr(S_1^c \cap S_2^c \cap \cdots S_{10}^c).$$

We may expand this as

$$\Pr(S_1^c \cap S_2^c \cap \cdots \cap S_{10}^c)$$

$$= 1 - \Pr(S_1 \cup S_2 \cup \cdots \cup S_{10})$$

$$= 1 - \left[\sum_{1 \le j_1 \le 10} \Pr(S_{j_1}) - \sum_{1 \le j_1 \le j_2 \le 10} \Pr(S_{j_1} \cap S_{j_2}) \right.$$

$$+ \sum_{1 \le j_1 \le j_2 \le j_3 \le 10} \Pr(S_{j_1} \cap S_{j_2} \cap S_{j_3}) - \cdots$$

$$\left. - \Pr(S_1 \cap S_2 \cap \cdots \cap S_{10}) \right].$$

But $\Pr(S_{j_1}) = \frac{9!}{10!}$ and $\Pr(S_{j_1} \cap S_{j_2}) = \frac{8!}{10!}$ and so forth. Since the number of ways to choose k indices from 10 is $\binom{10}{k}$, the above becomes

$\Pr(S_1^c \cap S_2^c \cap \cdots S_{10}^c)$

$$= 1 - \left[\binom{10}{1} \frac{9!}{10!} - \binom{10}{2} \frac{8!}{10!} + \binom{10}{3} \frac{7!}{10!} - \cdots - \binom{10}{10} \frac{1!}{10!} \right]$$

$$= 1 - \left[\frac{1}{1!} - \frac{1}{2!} + \frac{1}{3!} - \cdots - \frac{1}{10!} \right]$$

$$= \frac{1}{1!} + \frac{1}{2!} - \frac{1}{3!} + \cdots + \frac{1}{10!} \approx .368.$$

Look under Derangements or Probability in the Index for similar problems.

S1987-5

A geometric construction: We give a construction that is an induction on the dimension. This construction and plausibility argument can be made into a proof, but the analytic proof below is more concise. For me, the idea comes from the inductive step from dimension 2 to dimension 3, but I will begin here with dimension 1, where the logic begins. Clearly, the two most distant points in the unit ball of dimension 1, i.e., the unit interval, are the points $+1$ and -1.

The three most distant points in the unit ball in dimension 2, i.e., the unit disk, are the vertices of any equilateral triangle with vertices on the unit circle. Indeed, if A, B, and C are points of the disk that do not satisfy this condition, then assuming A and B are on the circle and the length of AC is least, then the triple A, B, C' is an improvement, where C' is the more distant intersection of the perpendicular bisector of AB and the circle. To find the coordinates of such points, we choose the north pole $(0,1)$ as the first point and note that the center of gravity of the three points must be the origin. Thus, the other two points, being symmetrically placed, have coordinates $(?, -1/2)$. Since they are on the circle, and at opposite ends of the segment (the 1-dimensional case), the coordinates must be $(\pm\sqrt{3}/2, -1/2)$.

The four most distant points in the unit ball in dimension 3 are four equidistant points on the sphere, (As before, if not, assuming A, B, and C are on the sphere and and the length of AD is least, then the 4-tuple A, B, C, D' is an improvement, where D' is the more distant intersection of the sphere and the line perpendicular to the disk determined by A, B, and C and through the center.) Again assuming the north pole $(0,0,1)$ to be one of the points, we find that the other three points must have coordinates $(?,?,-1/3)$, Moreover, these three points solve the problem for the disk they determine, that is, they are a scaled version of the solution of the 2-dimensional problem. To be on the sphere, since the third coordinate is $-1/3$, the scale factor must be $\sqrt{8/9}$ so the three points are $(0,\sqrt{8/9},-1/3)$ and $(\pm\sqrt{2/3},-\sqrt{2/9},-1/3)$.

The solution to the stated problem is five equidistant points on the unit sphere in dimension 4. Assuming the north pole is one of the points, the others have coordinates $(?,?,?,-1/4)$. The first three coordinates must be a scaled version of the solution to the 3-dimensional problem, and the scale factor must be $\sqrt{15/16}$. Thus,the solution to the problem is

$$(0,0,0,1); \quad (0,0,\sqrt{15/16},-1/4); \quad (0,\sqrt{5/6},-\sqrt{5/48},-1/4);$$

$$(\pm\sqrt{5/8},-\sqrt{5/24},-\sqrt{5/48},-1/4)$$

and the mutual distance is $\sqrt{5/2}$.

An analytic proof: Letting $v_1,\ldots,v_5$ be the five vectors above, we see that, for all i,j,

$$\|v_i - v_j\|^2 = 5/2.$$

Now, if $x_1,\ldots,x_5$ are any vectors in the unit ball of R^4, then

$$S = \sum_{i\neq j} \|x_i - x_j\|^2 = \sum_{i\neq j} \|x_i\|^2 + \|x_j\|^2 - 2x_i x_j$$

$$\leq 40 - 2\sum_{i\neq j} x_i x_j.$$

But

$$0 \leq \left(\sum_{i=1}^{5} x_i\right)\cdot\left(\sum_{i=1}^{5} x_i\right) = 5 + \sum_{i\neq j} x_i x_j,$$

so

$$-\sum_{i\neq j} x_i x_j \leq 5.$$

Combining this with the earlier inequality, we find

$$S \leq 40 + 10 = 50.$$

Since the sum for S contains 20 terms, either all of the terms are 2.5 or there is at least one of the terms that is less than 2.5. That is, the minimum distance between points is no more than $\sqrt{5/2}$ and the above solution is optimal.

Look under Geometry in the Index for similar problems.

S1987-6

If Jane were flying straight north, she would fly a quarter of the way around the earth before reaching the north pole, that is, she would fly $21\pi(4000)/4 = 2000\pi$ miles. Instead, her path is northwest, so she flies further. To be precise, at each instant, her path is along the hypotenuse of a right triangle with north-south and east-west legs and, since the surface of a sphere is locally like the Euclidean plane, the distance along the hypotenuse is $\sqrt{2}$ times the distance along the north-south leg. Since this is true at every instant of her journey, the total distance she flies before reaching the north pole is $2000\sqrt{2}\pi$ miles.

To see how many times she crosses the Greenwich Meridian, we look at the same infinitesimal triangle and relate the latitude and longitude angles. In the usual spherical coordinates with the origin at the center of the earth, longitude corresponds to the angle θ and (north) latitude corresponds to $\pi/2 - \varphi$. Thus, we want the total change in θ when φ changes from $\pi/2$ to 0, that is, we want

$$\left| \int_{\pi/2}^{0} d\theta(\varphi) \right|.$$

The north-south leg of the triangle has length $4000\,d\varphi$ whereas the east-west leg of the triangle has length $4000 \sin \varphi d\theta$ since the parallel corresponding to φ is a circle of radius $4000 \sin \varphi$. Thus, the total change in θ is then

$$\left| \int_{\pi/2}^{0} \frac{1}{\sin \varphi} d(\varphi) \right| = \infty.$$

which means Jane crosses the Greenwich Meridian infinitely often.

Look under Integration in the Index for similar problems.

Exam #23–1988

S1988-1

Let $x(t)$ denote distance between x's car and the finish line, and $y(t)$ denote the distance between y's car and the finish line. Suppose $x(0) = 2$

and $y(0) = 5$. For $t > 0$,

$$\frac{d^2x}{dt^2} = a(\frac{dx}{dt})^2 \Rightarrow \frac{dx}{dt} = \frac{-1}{at+b} \Rightarrow x(t) = \frac{-\ln(at+b)}{a} + c.$$

But

$$x(0) = 2 \Rightarrow c = 2 + \frac{\ln b}{a} \Rightarrow x(t) = 2 + \frac{\ln(\frac{b}{b+at})}{a}.$$

Let $t_1 > 0$ be such that $x(t_1) = 1$ and, hence,

$$\left.\frac{dx}{dt}\right|_{t_1} = \frac{1}{2}\left.\frac{dx}{dt}\right|_0 .$$

We have

$$1 = 2 + \frac{\ln(\frac{b}{b+at_1})}{a} \quad \text{and} \quad \frac{-1}{at_1+b} = \frac{1}{2}(\frac{-1}{b}),$$

implying that $a = \ln 2$.

Let t_x be the time required for driver x to reach the finish line. Then $0 = x(t_x)$ and $a = \ln 2$ together imply $t_x = \frac{3b}{\ln 2}$. Let t_y be the time required for driver y to reach the finish line. Then

$$\left.\frac{dy}{dt}\right|_{\text{all } t} = \left.\frac{dx}{dt}\right|_{t=0} = \frac{-1}{b}$$

and $y(0) = 5$ together imply $y(t) = 5 - \frac{t}{b}$, so that $t_y = 5b$.

Finally,

$$\ln 2 > \frac{3}{5} \Rightarrow t_y = 5b > \frac{3b}{\ln 2} = t_x,$$

since

$$0 > \left.\frac{dx}{dt}\right|_{t=0} = \frac{-1}{b}$$

implies $b > 0$. Driver x finishes first.

Look under Differential Equations in the Index for similar problems.

S1988-2

Let x and y denote any two members of the group. Then

$$x * x^{-1} * y = e * y = y = y * e = y * x^{-1} * x \Rightarrow x * y = y * x,$$

by the given cancellation rule.

Look under Group Theory in the Index for similar problems.

S1988-3

Note that

$$P(x) = 5 \cdot \sum_{k=1}^{n} T_k(x) = 5 \cdot \frac{d}{dx} \left[\prod_{i=1}^{n} (x - i) \right].$$

So if s and t are as prescribed,

$$\int_s^t P(x)\, dx = 5 \cdot \left[\prod_{i=1}^{n} (x - i) \right] \Bigg|_{x=s}^{x=t} = 0.$$

S1988-4

The unit's digit of $S_{12345} = 5$. To verify this it suffices to show, using induction on n, that $S_{5n} = 5n + 10k_n$ where k_n is a positive integer for each n. $S_5 = 1^2 + 3^2 + 5^2 + 7^2 + 9^2 = 165 = 5 + 10(16)$. Assuming the result true for n gives

$$
\begin{aligned}
S_{5(n+1)} &= S_{5n+5} \\
&= S_{5n} + [2(5n+1) - 1]^2 + \cdots + [2(5n+5) - 1]^2 \\
&= S_{5n} + (10n+1)^2 + (10n+3)^2 + \cdots + (10n+9)^2 \\
&= S_{5n} + 10j_n + (1^2 + 3^2 + \cdots + 9^2) \\
&= 5n + 10k_n + 10j_n + 165 \\
&= 5(n+1) + 10k_{n+1},
\end{aligned}
$$

where $k_{n+1} = k_n + j_n + 16$.

Look under Number Theory or Sequences in the Index for similar problems.

S1988-5

The desired area is 8 times that of the shaded region S in shown below. If the circle has radius a, then a Cartesian equation for the curve bounding S on the right is $(x + a)^2 + y^2 = 2a^2$. Changing to polar coordinates gives the polar equation $r = f(\theta) = a[\sqrt{1 + \cos^2 \theta} - \cos \theta]$. So the area of S is

$$\int_0^{\frac{\pi}{4}} \frac{f(\theta)^2}{2}\, d\theta = \frac{a^2}{2} \left[\frac{1 - \sqrt{3}}{2} + \frac{\pi}{3} \right]$$

after much work. Hence, the desired area is

$$2a^2 \left[1 - \sqrt{3} + \frac{\pi}{6} \right].$$

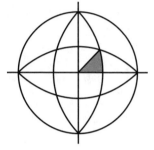

Look under Geometry or Integration in the Index for similar problems.

S1988-6

First draw the two straight lines BAP and CDP, where B and D are opposite ends of the diameter lying within L. Extend straight lines AD and BC downward until they intersect in a point E. PE is then a straight line perpendicular to L.

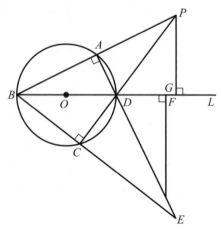

To prove this, notice that line segments EA and PC are altitudes of $\triangle BPE$. Since these segments are concurrent with the line segment through BD, that line segment must also be an altitude of $\triangle BPE$. Thus PE is perpendicular to line L.

Look under Analytic Geometry or Geometry in the Index for similar problems.

S1988-7

There are $2^{10} = 1024$ possible 10-flip sequences. Let H_n denote the number of n-flips sequences not having two or more tails in a row but ending with heads and let T_n denote the number of n-flips sequences not having two or more tails in a row but ending with a tail. Clearly $H_2 = 2$ and $T_2 = H_1 = 1$. For $n \geq 3$, $T_n = H_{n-1}$ and $H_n = H_{n-1} + T_{n-1} = H_{n-1} + H_{n-2}$. So $H_3 = H_2 + H_1 = 2 + 1 = 3$, $H_4 = 3 + 2 = 5$, ..., $H_9 = 55$, and $H_{10} = 89$. We need $H_{10} + T_{10} = 89 + 55 = 144$ and the desired probability is $\frac{144}{1024} = \frac{9}{64}$.

Look under Probability in the Index for similar problems.

Exam #24–1989

S1989-1

It is easy to show that there are three tests and that $x + y + z = 13$. From C's total we deduce that $3z \leq 9$, so that $z = 1$, 2, or 3. The only solution satisfying all the constraints of the problem is $z = 1$, $y = 4, x = 8$. It follows that A scored $4 + 8 + 8$, B scored $8 + 1 + 1$, and C scored $1 + 4 + 4$. Thus C came in second on the Geometry test.

Look under Logic in the Index for similar problems.

S1989-2

Any constant sequence satisfies the hypothesis of the problem. Conversely, if we fix m and let n tend to infinity, we see that $\lim_{n \to \infty} x_n = x_m$. By uniqueness of limits, we deduce that $x_1 = x_2 = \ldots$, so that $\{x_1, x_2, \ldots\}$ must be a constant sequence.

Look under Sequences in the Index for similar problems.

S1989-3:

If we label the figure as shown below and compare areas, we deduce that

$$2a = d,$$

$$2(a + h) = d + g + j,$$

$$2(a + h + e + b) = d + g + j + c + f + i.$$

These equations may be reduced to

$$2a = d,$$

$$2h = g + j,$$

$$2(e + b) = c + f + i.$$

In a similar fashion we have

$$2b = e,$$

$$2i = h + j,$$

$$2(f + c0 = a + d + g,$$

and

$$2c = f,$$

$$2g = i + j,$$

$$2(d + a) = b + e + h.$$

From these nine equations we deduce that

$$6j = a + b + c + d + e + f + g + h + i,$$

and this is equivalent to the assertion of the problem.

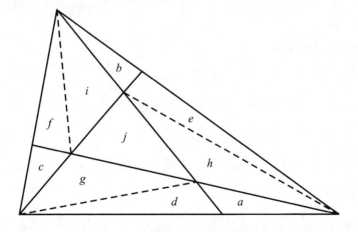

Look under Geometry in the Index for similar problems.

S1989-4

Assuming independence, if the fair coin is the one flipped only once, then the probability of obtaining the results noted is

$$\frac{1}{2} \cdot \binom{3}{2} \left(\frac{2}{3}\right)^2 \left(\frac{1}{3}\right)^1 = \frac{2}{9}.$$

On the other hand, if the biased coin is the one flipped only once, then the probability of obtaining the results noted is

$$\frac{2}{3} \cdot \binom{3}{2} \left(\frac{1}{2}\right)^2 \left(\frac{1}{2}\right)^1 = \frac{1}{4}.$$

Hence it is more likely that the biased coin was the one flipped only once.

Look under Probability in the Index for similar problems.

S1989-5

Let the y- and z-axes be the axes of the cylinders. If we slice the intersection by planes parallel to the yz-plane, the cross-sectional areas are all squares (by symmetry). Hence the volume is

$$\int_{-a}^{a} 4b^2 \left(1 - \frac{x^2}{a^2}\right) dx = \frac{16ab^2}{3}.$$

Look under Volumes in the Index for similar problems.

S1989-6

The characteristic equation is

$$(1 - \lambda)^n + (x_1 y_1 + \cdots + x_n y_n)(1 - \lambda)^{n-1} = 0.$$

Hence the eigenvalues are $\lambda = 1$ (multiplicity $n - 1$) and

$$\lambda = 1 + x_1 y_1 + \cdots + x_n y_n.$$

The determinant, being the product of the eigenvalues, is

$$1 + x_1 y_1 + \cdots + x_n y_n.$$

Look under Integration or Matrix Algebra in the Index for similar problems.

S1989-7

The dimensions are 4×18. A complete solution requires lengthy discussion to eliminate many potential solutions. A key ingredient is the fact that the number 72 has three "admissible" factorizations: 4×18, 6×12, and 8×9 with answers YES, NO, NO to the question "is the short side strictly less than half the long side?" Several other possible areas admit triple factorizations, e.g., $48 = 3 \times 16$, 4×12 and 6×8, but here the answers are always two YESes and one NO.

Note: The students were told at the beginning of the contest to assume that the sides of Smith's ranch are parallel to the side of Todd county.

Look under Logic in the Index for similar problems.

Exam #25–1990

1990-1

Write $n = 10^k a_k + 10^{k-1} a_{k-1} + \cdots + 10a_1 + a_0$, where $a_i \in \{0, \ldots, 9\}$ and $a_k \geq 1$. We assume

$$10^k a_k + 10^{k-1} a_{k-1} + \cdots + 10a_1 + a_0 = a_k^2 + \cdots + a_0^2 + 1$$

which implies

$$(10^k - a_k)a_k + \cdots + (10 - a_1)a_1 = a_0^2 - a_0 + 1. \qquad (*)$$

The right-hand side of (*) is at most $9^2 - 9 + 1 = 73$. Since $a_k \geq 1$, the left-hand side of (*) is at least $10^k - 9$. Therefore $k \leq 1$. Clearly, there are no 1 digit solutions. It is easy to check that the only solutions to $(10 - a_1)a_1 = a_0^2 - a_0 + 1$ are $a_0 = 5$, $a_1 = 3$ or 7, so 35 or 75 are the only solutions.

Look under Number Theory in the Index for similar problems.

1990-2

This is an inclusion-exclusion problem, where the universe U is the set of arrangements of I, N, D, I, A, N, A and the subsets $A_1, \ldots, A_6$ are those arrangements in which each possible digraph IA, AI, IN, NI, AN, NA, is repeated. If, say, IA is repeated, then neither IN, nor NA can occur at all, and AI could occur at most once. However AN or NA might occur twice if the tuples IA or NIA were repeated. In other words,

$$|A_1 \cap A_2| = |A_1 \cap A_3| = |A_1 \cap A_6| = 0.$$

Continuing in this way, we find that 6 of the pairs $A_i \cap A_j$ are possible and none of the 20 tuples $A_i \cap A_j \cap A_k$ may occur simultaneously. Thus the total number of possibilities (by symmetry) is

$$|U| - 6|A_i| + 6|A_i \cap A_j|$$

assuming $A_i \cap A_j$ is possible.

Note that $|U| = 7!/[(2!)^3 \cdot 1!] = 630$. If IA is repeated, then we wish to write (IA), (IA), N, D, N in some order which can be done in $5!/[(2!)^2 \cdot 1!] = 30$ ways. If IA and AN are repeated, then we need to arrange (IAN),(IAN), D, which can be done in $3!/(2! \cdot 1!) = 3$ ways. Thus the desired answer is $630 - 6 \cdot 30 + 6 \cdot 3 = 468$.

Look under Enumeration in the Index for similar problems.

S1990-3

Since $f'(x) = f(x+1) - 2$, f' is differentiable and

$$f''(x) = f'(x+1) = f(x+2) - 2 = f(x) - 2.$$

It follows that $f(x) = 2 + c_1 e^x + c_2 e^{-x}$, and, since $f(x+2) = f(x)$, $c_1 e^2 = c_1$ and $c_2 e^{-2} = c_2$ so $c_1 = c_2 = 0$.

S1990-4

A variety of approaches work. This one refers to the figure below. Since the exterior angle to a regular heptagon has measure $\frac{2\pi}{7}$, $\angle GAB = \frac{5\pi}{7}$. Since $\angle GAF$ subtends an arc of $\frac{\pi}{7}$ on the circle, its measure is $\frac{\pi}{7} = \angle GAK = \angle LAB$, and $\angle KAL = \frac{3\pi}{7}$, so $\angle AKL = \angle ALK = \frac{2\pi}{7}$. The law of sines applied to r and to s and l gives

$$\frac{r}{\sin \frac{3\pi}{7}} = \frac{s}{\sin \frac{2\pi}{7}} \quad \text{and} \quad \frac{s}{\sin \frac{\pi}{7}} = \frac{1}{\sin \frac{5\pi}{7}},$$

so

$$r = \frac{\sin \frac{\pi}{7} \sin \frac{3\pi}{7}}{\sin \frac{2\pi}{7} \sin \frac{5\pi}{7}}.$$

This may be reduced to (among others):

$$\frac{\sin \frac{\pi}{7} \sin \frac{3\pi}{7}}{\sin^2 \frac{2\pi}{7}}, \quad 2\cos \frac{\pi}{7} - \frac{1}{\cos \frac{\pi}{7}}, \quad \frac{\cos \frac{2\pi}{7}}{\cos \frac{\pi}{7}}.$$

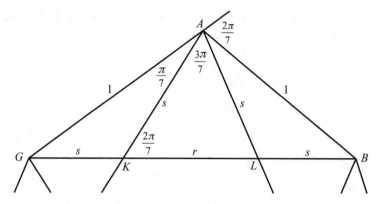

Other proofs use the law of cosines, and dropping a perpendicular from K to AG, etc.

Look under Geometry in the Index for similar problems.

S1990-5

Construct the Riemann sum

$$n^{-2}\sum_{i=1}^{n}\sum_{j=1}^{n^2}\frac{1}{\sqrt{n^2+ni+j}}=\sum_{i=1}^{n}\sum_{j=1}^{n^2}\frac{1}{n}\cdot\frac{1}{n^2}\frac{1}{\sqrt{1+i/n+j/n^2}},$$

which in the limit, gives us

$$\int_0^1\int_0^1\frac{dx\,dy}{\sqrt{1+x+y}}=\int_0^1\left(2\sqrt{1+x+y}\right)\big|_{y=0}^{1}\,dx$$

$$=\int_0^1\left(2\sqrt{2+x}-2\sqrt{1+x}\right)dx$$

$$=\frac{4}{3}(3^{3/2}-2\cdot2^{3/2}+1^{3/2})=\frac{4}{3}(3\sqrt{3}-4\sqrt{2}+1)$$

$$(\approx.719).$$

Look under Limit Evaluation or Riemann Sums in the Index for similar problems.

S1990-6

Insert a coordinate system centered at the center of the hexagon, so the vertices of the hexagon are $(0,\pm1)$, $(\pm\frac{\sqrt{3}}{2},\pm\frac{1}{2})$. The parabolas must have the form $y=\alpha x^2+\beta$, and the points show that they are $y=1-\frac{2}{3}x^2$ and $y=-1+\frac{2}{3}x^2$, which intersect at $(\pm\frac{\sqrt{3}}{2},0)$. A routine calculation shows that the area between the parabolas is $\frac{1}{3}\sqrt{6}$ and that of the hexagon is $6 \cdot \frac{\sqrt{3}}{4}$, so the total area of the shaded region is

$$\frac{8\sqrt{2} - 9}{2\sqrt{3}} = \frac{8\sqrt{6} - 9\sqrt{3}}{6}.$$

Look under Analytic Geometry in the Index for similar problems.

Exam #26–1991

S1991-1

The curve, clearly symmetric with respect to both the x-axis and the y-axis, lies on or between the lines $x = \pm 1$. On $[-1, -1/2]$, $y = \pm\sqrt{3}(x+1)$. On $[-1/2, 1/2]$, $y = \sqrt{3}/2$. On $[1/2, 1]$, $y = \sqrt{3}(1-x)$. The curve is a regular hexagon of edge length 1 with vertices $(\pm 1/2, \sqrt{3}/2)$, $(\pm 1/2, \sqrt{3}/2)$ and $(\pm 1, 0)$. The enclosed area (that of six equilateral triangles of unit edge length) is $6 \times 1/2 \times 1 \times 1 \times \sqrt{3}/2 = 3\sqrt{3}/2$.

Look under Analytic Geometry in the Index for similar problems.

S1991-2

The number of k-digit numbers without any 1's is $8 \times 9^{k-1}$, so between 1 and 1,000,000 inclusive there are $8(1 + 9 + 9^2 + \cdots + 9^5) = 9^6 - 1 = 531,440$ numbers without 1's and 468,560 numbers with 1's.

Alternate Solution   The number of ways of filling six spaces with digits 0, 2, 3, 4, 5, 6, 7, 8, 9 is 9^6. But 000000 is not between 1 and 1,000,000, so there are $9^6 - 1 = 531,440$ numbers without 1's and 468,560 with 1's.

Look under Enumeration in the Index for similar problems.

S1991-3

Label the areas of the eleven regions inside the parallelogram as illustrated below.

Let K be the area of the parallelogram $ABCD$; then

(1) area $ABR = K/2 = r_3 + w_2 + w_3 + w_5 + b$,

(2) area ARD + area $BRC = K/2 = r_2 + w_1 + r_1 + r_4 + w_4 + r_5$,

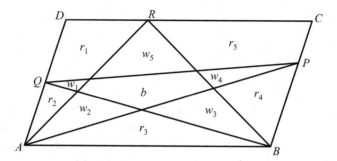

(3) area QAB = area QAP, so $r_3 = w_1 + b + w_4$.

(4) From (1) and (2), $r_3 + w_2 + w_3 + w_5 + b = r_2 + w_1 + r_1 + r_4 + w_4 + r_5$.

(5) Substituting $r_3 = w_1 + b + w_4$ from (3) in the left side of (4) and $w_1 + w_4 = r3 - b$ (from (3)) in the right side, $\sum w_i + 2b = \sum r_i - b$.

Using K as the sum of the reds + sum of the "whites" + blue, we get red area less the blue area is $K/2$ regardless of the choices of points P, Q, and R.

Look under Enumeration or Geometry in the Index for similar problems.

S1991-4

The sum $a + b + c$ is even, so the sum $-a + b + c$ is even since parity is not affected by a sign change in a summand. The identity

$$a^2 + b^2 + c^2 - 2ab - 2bc - 2ca = a^2 + b^2 + c^2 - 2ab + 2bc - 2ca - 4bc$$

suggests taking

$$n = \frac{a^2 + b^2 + c^2 - 2ab - 2bc - 2ca}{4}.$$

Indeed,

$$\left(\frac{-a + b + c}{2} \right)^2 = n + bc,$$

$$\left(\frac{a - b + c}{2} \right)^2 = n + ba, \qquad \text{and}$$

$$\left(\frac{a + b - c}{2} \right)^2 = n + ab.$$

Look under Number Theory in the Index for similar problems.

S1991-5

Remark: If $n = a^2 + b^2$ and $m = c^2 + d^2$, then $nm = (ac - bd)^2 + (ad + bc)^2$. Thus, the product of two sums of squares is also a sum of squares.

(a) Take $a = c$ and $b = d$, then $(a^2 - b^2)^2 + (2ab)^2 = (a^2 + b^2)^2$. It follows that if n is a sum of two squares, then n^2 is a sum of two squares. The first example $1^2 + 4^2 = 17$, $3^2 + 3^2 = 18$ leads to infinitely many more, namely $8^2 + 15^2 = 17^2 = 289$, $1^2 + 17^2 = 290$, and so on. From the example $3^2 + 4^2 = 5^2 = 25$, $1^2 + 5^2 = 26$, one can generate infinitely many more by taking $(3k)^2 + (4k)^2 = (5k)^2 = 25k^2$ and $1^2 + (5k)^2 = 1 + 25k^2$. Any Pythagorean triple will serve as a starting point.

(b) If $n - 1$, n, $n + 1$ are consecutive sums of two squares, then $n^2 - 1 = (n - 1)(n + 1)$, n^2, $n^2 + 1$ are consecutive sums of two squares. For starters, take $72 = 6^2 + 6^2$, $73 = 3^2 + 8^2$, $74 = 5^2 + 7^2$.

(c) In any set of four consecutive positive integers one must be congruent to 3 modulo 4. This integer cannot be a sum of squares since $a^2 + b^2 \equiv 0, 1$ or $2 \bmod 4$ for any integers a and b. Therefore, it is not possible to have four consecutive positive integers which are sums of two squares.

Look under Number Theory in the Index for similar problems.

S1991-6

We have

$$\lim_{n \to \infty} \frac{n}{H_n} = \lim_{n \to \infty} n \left(\frac{1}{n} \right) \left(\frac{1}{n+1} + \frac{1}{n+2} + \cdots + \frac{1}{n+n} \right)$$

$$= \lim_{n \to \infty} \left(\frac{1}{n} \right) \left(\frac{1}{1 + 1/n} + \frac{1}{1 + 2/n} + \cdots + \frac{1}{1 + n/n} \right)$$

$$= \lim_{n \to \infty} \left(\frac{1}{n} \right) \sum_{k=1}^{n} \frac{1}{1 + k/n}$$

$$= \int_0^1 \frac{1}{1 + x}\, dx$$

$$= \ln(1 + x)\big|_0^1 = \ln 2.$$

Therefore,

$$\lim_{n \to \infty} \frac{H_n}{n} = \frac{1}{\ln 2}.$$

Look under Limit Evaluation in the Index for similar problems.

S1991-7

By reducing the coordinates of each of the nine given points modulo 3, each can be identified with one of the nine points in the square array S:
$$\{(-1,-1),(-1,0),(-1,1),(0,-1),(0,0),(0,1),(1,-1),(1,0),(1,1)\}.$$
The existence of three lattice points with a lattice point as centroid is equivalent to having a row, column, or diagonal (broken diagonals included) of a 3×3 square array completely occupied. If only four or fewer squares were occupied by the nine points, then one or more would be occupied by at least three points which would have a lattice point as centroid. Therefore, at least five of the squares in the 3×3 square array must be occupied. To avoid three points in a row, two rows would contain just two points each. It is now easy to show that it is impossible to place the fifth without completing a row, column, or diagonal, (possibly broken).

Look under Analytic Geometry or Enumeration in the Index for similar problems.

Exam #27–1992

S1992-1

The curve is a cycloid which can be parameterized as $x = t - \sin t$ and $y = 1 - \cos t$ for $0 \leq t \leq 2\pi$. The arc length is given by

$$\int_0^{2\pi} \sqrt{(1-\cos(t))^2 + (\sin(t))^2}\, dt = \int_0^{2\pi} \sqrt{2(1-\cos(t))}\, dt$$

$$= \int_0^{2\pi} \sqrt{4\sin^2(t/2)}\, dt$$

$$= -4\cos(t/2)\Big|_0^{2\pi} = 8.$$

Look under Integration in the Index for similar problems.

S1992-2

Observe that the ith tallest person is viewable if and only if he is in front of all the people $1, 2, 3, \ldots, i-1$. This happens with probability $\frac{1}{i}$. Therefore, the answer is

$$\sum_{i=1}^{n} \frac{1}{n} \approx \ln n + 0.5772157\ldots.$$

Look under Probability in the Index for similar problems.

S1992-3

Select a head gossip v_0. First have all $n-1$ other people send a letter with their scandal to v_0; then have v_0 distribute the assembled information via $n-1$ letters to everyone else. We claim that this total of $2n-2$ letters is the minimum.

Let each letter be represented by an arc in a directed graph. No one can know all the scandal until at least $n-1$ letters have been sent, for we need a subgraph of arcs unilaterally connected with directed paths from every vertex into whichever vertex becomes fully informed. Then each subsequent letter can inform one more person, so at least $n-1$ more letters are required. This gives a bound of $2n-2$ which we have already achieved above.

Look under Enumeration in the Index for similar problems.

S1992-4

For $n=1$, Z_1 has no proper subgroup, contrary to the stated condition. For $n=2$, Z_2 works because it has 1 subgroup. For $n=3$, Z_3 fails because it has 1, and not 2, proper subgroups. For $n=4$, we need 4 subgroups. The Klein 4-group works, whereas Z_4 only has 2. Thus we have found 2 groups that satisfy this condition, namely Z_2 and K_4.

We shall now show that no group of order $n \geq 5$ has so many subgroups. First, every subgroup contains the identity element, e, so the number of subgroups of order i cannot exceed the number of subsets of order $i-1$ selected from $G-\{e\}$. There are $\binom{n-1}{i-1}$ of these, and the largest subgroup order i is at most $\lfloor \frac{n}{2} \rfloor$. For any odd $n=2k+1$, this yields at most

$$\sum_{i=1}^{k} \binom{n-1}{i-1} < 2^{2k-1}$$
$$= 2^{n-2}$$

subgroups. For any even $n=2k$, we can find at most

$$\sum_{i=1}^{k} \binom{n-1}{i-1} = 2^{2k-1}$$
$$= 2^{n-2}$$

subgoups. That is just barely enough, but only if every set of order $\lfloor \frac{n}{2} \rfloor$ is a subgroup. For $n \leq 4$, this is true of Z_2 and K_4.

But for $n \geq 6$ this requires every subset of order 2 and 3 be a subgroup. That is, $\{e, a\}$ and $\{e, a, b\}$ must both be subgroups. But the first forces the order of a to be 2 while the second requires the order of a to be 3, a contradiction.

Look under Group Theory in the Index for similar problems.

SI992-5

Set $I = \int_{-\infty}^{\infty} e^{-x^2}\, dx = \int_{-\infty}^{\infty} e^{-y^2}\, dy$. Then

$$
\begin{aligned}
I^2 &= \int_{-\infty}^{\infty} e^{-x^2}\, dx \cdot \int_{-\infty}^{\infty} e^{-y^2}\, dy \\
&= \int_{-\infty}^{\infty} \int_{-\infty}^{\infty} e^{-x^2-y^2}\, dy\, dx \\
&= \int_{0}^{\infty} \int_{0}^{2\pi} e^{-r^2} r\, d\theta\, dr \\
&= \int_{0}^{\infty} e^{-r^2} 2\pi r\, dr \\
&= -\pi e^{-r^2}\big|_0^\infty = \pi.
\end{aligned}
$$

Therefore $I = \sqrt{\pi}$.

Look under Integration in the Index for similar problems.

SI992-6

(a) The characteristic polynomial of A is

$$
X^2 - 13X + 36 = (X - 4)(X - 9).
$$

We find that the eigenvectors for A are $\begin{pmatrix} -1 \\ 3 \end{pmatrix}$ and $\begin{pmatrix} 1 \\ -2 \end{pmatrix}$. This gives

$$
P = \begin{pmatrix} -1 & 1 \\ 3 & -2 \end{pmatrix} \quad \text{and} \quad P^{-1} = \begin{pmatrix} 2 & 1 \\ 3 & 1 \end{pmatrix},
$$

and therefore

$$
A = \begin{pmatrix} -1 & 1 \\ 3 & -2 \end{pmatrix} \begin{pmatrix} 4 & 0 \\ 0 & 9 \end{pmatrix} \begin{pmatrix} 2 & 1 \\ 3 & 1 \end{pmatrix}.
$$

(b) Clearly all four choices of signs in $C = \begin{pmatrix} \pm 2 & 0 \\ 0 & \pm 3 \end{pmatrix}$ yield $C^2 = \begin{pmatrix} 4 & 0 \\ 0 & 9 \end{pmatrix}$. Therefore, all four choices of $B = PCP^{-1}$ give

$$
B^2 = PCP^{-1}PCP^{-1} = PC^2P^{-1} = A
$$

as desired. These are

$$B = \begin{pmatrix} 5 & 1 \\ -6 & 0 \end{pmatrix}, \begin{pmatrix} -5 & -1 \\ 6 & 0 \end{pmatrix}, \begin{pmatrix} 13 & 5 \\ -30 & -12 \end{pmatrix}, \begin{pmatrix} -13 & -15 \\ 30 & 12 \end{pmatrix}.$$

(c) If A has a repeated eigenvalue, for example if $A = 2^n I$, we can find at least $2(n + 1)$ different B's; namely, for any i with $0 \le i \le n$ we may have

$$B = \pm \begin{pmatrix} 0 & 2^i \\ 2^{n-i} & 0 \end{pmatrix}.$$

Thus, there is no maximum.

Look under Matrix Algebra in the Index for similar problems.

S1992-7

Solution 1 Since $y = x^2$ is a continuous function, we may square $x/\sqrt{1 - e^{-2x^2}}$ to get

$$\lim_{x \to 0+} \frac{x^2}{1 - e^{-2x^2}} = L^2.$$

Now this has the form $\frac{0}{0}$ so we may apply L'Hôpital's Rule to get

$$\lim_{x \to 0+} \frac{2x}{4xe^{-2x^2}} = L^2 = \frac{1}{2}.$$

Therefore $L = \pm(1/\sqrt{2})$. But $L \ge 0$ since the fraction is positive on the domain. Therefore $L = +(1/\sqrt{2})$.

Solution 2 Since the form is $\frac{0}{0}$, we apply L'Hôpital's Rule to get

$$\lim_{x \to 0+} \frac{\sqrt{1 - e^{-2x^2}}}{2xe^{-2x^2}}.$$

Since $e^{-2x^2} \to 1$, this reduces to

$$\lim_{x \to 0+} \frac{\sqrt{1 - e^{-2x^2}}}{2x},$$

provided this limit exists. But this also equals $L/2$ provided $L \ne 0$. We find $L = \frac{1}{2L}$ so that, as above, $L = \pm(1/\sqrt{2})$.

Look under Limit Evaluation in the Index for similar problems.

S1992-8

We want integer solutions to $a^2 + (a + 1)^2 = b^2$. This can be rewritten as

$$1 = 2b^2 - (2a + 1)^2.$$

Upon replacing $c = 2a + 1$, we have a Pell equation $1 = 2b^2 - c^2$. The first solution is $b_1 = c_1 = 1$. In general, given a solution (b_n, c_n), we can generate another solution

$$(b_{n+1}, c_{n+1}) = (3b_n + 2c_n, 4b_n + 3c_n)$$

because

$$2(3b_n + 2c_n)^2 - (4b_n + 3c_n)^2 = 2b_n^2 - c_n^2 = 1.$$

Thus, we find the solutions $(b_2, c_2) = (5, 7)$, $(b_3, c_3) = (29, 41)$, $(b_4, c_4) = (169, 239)$, $\ldots$.

Can there be any other solutions? We can also write the recurrence in reverse as $(b_n, c_n) = (3b_{n+1} - 2c_{n+1}, -4b_{n+1} + 3c_{n+1})$. Suppose that (b, c) is the smallest solution not on our indexed list.

Upon applying the reverse recurrence we find that $(3b - 2c, -4b + 3c)$ is another smaller solution that also cannot be in our list. But this contradicts our selection of (b, c). Consequently the list includes every possible solution.

Can we find an explicit solution of the simultaneous recurrence? Applying the recurrences repeatedly we find

$$\begin{aligned} b_{n+1} &= 3b_n + 2c_n \\ &= 3b_n + 2(4b_{n-1} + 3c_{n-1}) \\ &= 3b_n + 8b_{n-1} + 6(-4b_n + 3c_n) \\ &= -21b_n + 8b_{n-1} + 18c_n. \end{aligned}$$

Now nine times the first equation minus the last yields

$$8b_{n+1} = 48b_n - 8b_{n-1} \quad \text{or} \quad b_{n+1} = 6b_n - b_{n-1}.$$

Similarly, we can show that c_n satisfies the same recurrence. Namely $c_{n+1} = 6c_n - c_{n-1}$. The roots of $x^2 - 6x + 1$ are associated with this recurrence, namely $r = 3 \pm \sqrt{8}$.

We find that

$$b_n = \left(\frac{2 - \sqrt{2}}{4}\right)(3 + \sqrt{8})^n + \left(\frac{2 + \sqrt{2}}{4}\right)(3 - \sqrt{8})^n,$$

$$c_n = \left(\frac{\sqrt{2} - 1}{2}\right)(3 + \sqrt{8})^n - \left(\frac{\sqrt{2} + 1}{2}\right)(3 - \sqrt{8})^n,$$

and recall

$$a_n = \frac{c_{n-1}}{2} = \left(\frac{\sqrt{2}-1}{4}\right)(3+\sqrt{8})^n - \left(\frac{\sqrt{2}+1}{4}\right)(3-\sqrt{8})^n.$$

Look under Diophantine Equations in the Index for similar problems.

Exam #28–1993

S1993-1

Let $P(x) = ax^3 + bx^2 + cx + d$. Then $P'(x) = 3ax^2 + 2bx + c$; $P''(x) = 6ax + 2b$; and $P'''(x) = 6a$. The following inequalities are equivalent:

$$\frac{P'''(x)}{P'(x)} < \frac{1}{2}\cdot\left(\frac{P''(x)}{P'(x)}\right)^2,$$

$$2P'(x)P'''(x) < \left(P''(x)\right)^2,$$

$$2(3ax^2 + 2bx + c)(6a) < (6ax + 2b)^2,$$

$$36a^2x^2 + 24abx + 12ac < 36a^2x^2 + 24abx + 4b^2,$$

$$12ac < 4b^2,$$

$$(2b)^2 - 4(3a)c > 0.$$

Since $P'(x)$ has distinct zeros, the last inequality holds and the result follows.

Look under Polynomials in the Index for similar problems.

S1993-2

For simplicity of notation, we write f and g for $f(a)$ and $g(a)$ and omit parentheses.

$$f = gff = (fgf)ff = f(gff)f = fff$$
$$= ff(gff) = f(fgf)f = fgf$$
$$= g.$$

Look under Group Theory in the Index for similar problems.

S1993-3

Since A has rank 1, it has at least one nonzero row, r, and each of its rows is a multiple of r. Therefore there is a $1 \times n$ row vector X for which $A = X^T \cdot r$. Now, for any $p \times q$ and $q \times p$ matrices B and C, the trace of $B \cdot C$ equals the trace of $C \cdot B$. Therefore we have

$$1 = \text{ trace of } A = \text{ trace of } (r \cdot X^T) = r \cdot X^T.$$

(The last equality holds since $(r \cdot X^T)$ is a 1×1 matrix.) Thus,

$$A^2 = (X^T \cdot r) \cdot (X^T \cdot r) = X^T \cdot (r \cdot X^T) \cdot r = X^T \cdot r = A.$$

Look under Matrix Algebra in the Index for similar problems.

S1993-4

Let $O(n)$ be the number of orderly permutations of $\{1, 2, \ldots, n\}$. We will prove by induction that $O(n) = 2^{n-1}$. Since both 12 and 21 are orderly, $0(2) = 2^{2-1} = 2$. Now assume that $O(k) = 2^{k-1}$ and note that any orderly permutation of $\{1, 2, \ldots, k\}$ must begin with 1 or k. (Let j be the first number. Wherever the 1 is, 2 must follow it; 3 must follow the 2; and so on and inductively, j must follow the 1.) For any orderly permutation, P, of $\{1, 2, \ldots, n\}$ we create two orderly permutations of $\{1, 2, \ldots, (k+1)\}$ as follows:

1) place $(k + 1)$ at the begining of P, or

2) increase each number in P by 1 and place 1 at the beginning.

(For example, with $k = 4$ and $P = 1423$, we create 51423 and 12534.)

This process thus creates 2^k distinct orderly permutations of $\{1, 2, \ldots, (k + 1)\}$. Since, for any orderly permutation of $\{1, 2, \ldots, (k + 1)\}$, we can reverse the process and obtain the orderly permutation of $\{1, 2, \ldots, k\}$ which created it, we see that $O(k + 1) = 2^k$, completing our induction.

Look under Permutations in the Index for similar problems.

S1993-5

Solution 1 Since $a_{n+1}^2 = a_n^2 + a_n$, one can show by induction that

$$\frac{n}{2} - \sqrt{n} \le a_n \le \frac{n}{2},$$

so

$$\frac{a_n}{n} \to \frac{1}{2}.$$

Solution 2   Since $a_{n+1}^2 = a_n^2 + a_n$, one has

$$a_{n+1} - a_n = \sqrt{a_n^2 + a_n} - a_n$$

$$= \frac{a_n^2 + a_n - a_n^2}{\sqrt{a_n^2 + a_n} + a_n}$$

$$= \frac{a_n}{\sqrt{a_n^2 + a_n} + a_n}$$

$$= \frac{1}{\sqrt{1 + \frac{1}{a_n}} + 1}.$$

Next since $a_1 = 1$, the a_n's are increasing and the calculations above show that

$$a_{n+1} \geq a_n + \frac{1}{\sqrt{2} + 1}$$

so $a_n \to \infty$. Next, using $a_n \to \infty$, the same calculations show that

$$a_{n+1} - a_n \to \frac{1}{2}.$$

This, in turn, shows

$$\frac{a_n}{n} \to \frac{1}{2}.$$

Look under Limit Evaluation in the Index for similar problems.

S1993-6

First note that there are integers A_n and B_n for which

$$(\sqrt{3} \pm 1)^{2n} = (4 \pm 2\sqrt{3})^n = A_n \pm B_n \sqrt{3}$$

and further that

$$(2 \pm \sqrt{3})^n = (4 \pm 2\sqrt{3})^n / 2^n = (A_n / 2^n) \pm (B_n / 2^n)\sqrt{3}.$$

Thus, since $(A_n / 2^n)$ is an integer, A_n is divisible by 2^n. Note that

$$(\sqrt{3} + 1)^{2n} + (\sqrt{3} \cdot 1)^{2n} = (4 + 2\sqrt{3})^n + (4 \cdot 2\sqrt{3})^n = 2A_n$$

is divisible by 2^{n+1}. Finally, since $0 < (\sqrt{3} \cdot 1)^{2n} < 1$, it follows that

$$\lceil (\sqrt{3} + 1)^{2n} \rceil = (\sqrt{3} + 1)^{2n} + (\sqrt{3} \cdot 1)^{2n} = 2A_n.$$

Look under Number Theory in the Index for similar problems.

S1993-7

If there are only two distinct integers, $X < Y$, among A, B, C, D, E then there would be at most three distinct sums of pairs of them: $X + X < X + Y < Y + Y$. Also, if there are at least four distinct integers, $W < X < Y < Z$ among A, B, C, D, E then there would be at least five distinct sums of pairs of them: $W + X < W + Y < X + Y < X + Z < Y + Z$. Therefore there are exactly three distinct integers among A, B, C, D, E and we have the following six cases:

1) $A = B = C < D < E$,   2) $A = B < C = D < E$,

3) $A = B < C < D = E$,   4) $A < B = C = D < E$,

5) $A < B = C < D = E$,   6) $A < B < C = D = E$.

The first three cases are impossible since the smallest sum 401, is odd. In case (4), we must have

$$A + B = 401, \quad 2B = 546, \quad A + E = 691, \quad \text{and} \quad B + E = 836.$$

(Why?) This yields

$$A = 128, \quad B = C = D = 273, \quad E = 563.$$

Similarly, in case (5) we have

$$A + B = 401, \quad 2B = A + D = 546, \quad B + D = 791, \quad \text{and} \quad 2D = 836,$$

which yields

$$A = 128, \quad B = C = 273, \quad D = E = 418.$$

In case (6) we have

$$A + B = 401, \quad A + C = 546, \quad B + C = 691, \quad \text{and} \quad 2C = 836,$$

from which we obtain

$$A = 128, \quad B = 273, \quad C = D = E = 418.$$

In summary, there are three collections of integers satisfying the given conditions:

$$A = 128, \quad B = C = D = 273, \quad E = 563;$$

$$A = 128, \quad B = C = 273, \quad D = E = 418;$$

$$\text{and} \quad A = 128, \quad B = 273, \quad C = D = E = 418.$$

Look under Logic or Systems of Equations in the Index for similar problems.

S1993-8

Let the area of $abcd$ be one 'unit'. We will determine the area of $ABCD$ in units. Since $\frac{Dd}{dc} = \frac{5}{1} = 5$, and $\frac{Ad}{ad} = \frac{8}{2} = 4$, we see that $\triangle AdD$ contains $\frac{5\times 4}{2} = 10$ units. Similarly, since $\frac{Ba}{ba} = \frac{6}{1} = 6$ and $\frac{Aa}{ad} = \frac{6}{2} = 3$, $\triangle AaB$ contains $\frac{6\times 3}{2} = 9$ units. Noting that $\triangle AdD = \triangle CbB$ and $\triangle AaB = \triangle CcD$, we see that $ABCD$ contains $2 \times 10 + 2 \times 9 + 1 = 39$ units.

Finally, since area of $ABCD = 48$, we see that the area of $abcd$ is $\frac{16}{13}$.

Note: In general it can be shown that if 1, 5, 2, and 6 are replaced by w, x, y, and z, then the area of $abcd$ is

$$\frac{wy(w+x)(y+z)}{xz+(w+x)(y+z)}.$$

Look under Geometry in the Index for similar problems.

Exam #29–1994

S1994-1

Using L'Hôpital's Rule,

$$\lim_{x\to\infty} \frac{\int_0^x e^{t^2}\, dt}{e^{x^2}/x} = \lim_{x\to\infty} \frac{e^{x^2}}{\frac{x\cdot 2xe^{x^2}-e^{x^2}}{x^2}}$$

$$= \lim_{x\to\infty} \frac{x^2}{2x^2-1} = \frac{1}{2}.$$

Look under Limit Evaluation in the Index for similar problems.

S1994-2

We have

$$(1+\sqrt{2/n})^n = 1 + n\sqrt{2/n} + \frac{n(n-1)}{2}\cdot\frac{2}{n} + \cdots + \left(\frac{2}{n}\right)^{n/2}$$

$$= n + \sqrt{2n} + \cdots + \left(\frac{2}{n}\right)^{n/2} > n.$$

Now take nth roots.

Look under Inequalities in the Index for similar problems.

S1994-3

The inequality is

$$\ln \frac{x^2}{y^2} < \frac{x}{y} - \frac{y}{x}.$$

Let $f(p) = \ln p^2 - p + p^{-1}$. Then $f'(p) = -p^{-2}(p-1)^2 < 0$, for $p > 1$. Since $f(1) = 0$, it follows that $f(p) < 0$ when $p > 1$. This implies the inequality.

Look under Inequalities in the Index for similar problems.

S1994-4

If C is the largest and A the smallest angle of the triangle, then the law of cosines gives $\cos C = 1/8$ and $\cos A = 3/4$. Then $\cos 2A = 2\cos^2 A - 1 = 1/8$ and it follows that, since A and C are between 0 and π, $2A = C$.

Look under Geometry in the Index for similar problems.

S1994-5

Let the integers in non-decreasing order be a, b, c, d, and e. Then $abcde = a + b + c + d + e \le 5e$ so $abcd \le 5$.

There are five cases:

$abcd = 1$ implies $e = 4 + e$, impossible,

$abcd = 2$ implies $2e = 5 + e$, so $e = 5$,

$abcd = 3$ implies $3e = 6 + e$, so $e = 3$,

$abcd = 4$ implies $4e = 6 + e$ or $4e = 7 + e$; the first gives $e = 2$ and the second is impossible, and

$abcd = 5$ implies $5e = 8 + e$, but since $d = 5$, $e = 2$ is impossible.

The three solutions are thus $\{1, 1, 1, 2, 5\}$; $\{1, 1, 1, 3, 3\}$; and $\{1, 1, 2, 2, 2\}$.

Look under Number Theory in the Index for similar problems.

S1994-6

Suppose that the circle contains three distinct rational points. Then, substituting their coordinates in $x^2 + y^2 + Ax + By + C = O$ determines rational values for A, B, and C. But then the coordinates of the center, $(-A/2, -B/2)$, are both rational, a contradiction.

Look under Analytic Geometry in the Index for similar problems.

Exam #30–1995

S1995-1

Let $u = z_1 + \sqrt{z_1^2 - z_2^2}$, and $v = z_1 - \sqrt{z_1^2 - z_2^2}$. Then $u + v = 2z_1$, and $u - v = 2\sqrt{z_1^2 - z_2^2}$. By the Parallelogram Identity,

$$|u|^2 + |v|^2 = \frac{1}{2}(|u + v|^2 + |u - v|^2)$$
$$= 2|z_1|^2 + 2|z_1^2 - z_2^2|,$$

from above. But, $|uv| = |z_1^2 - (z_1^2 - z_2^2)| = |z_2|^2$. Hence

$$(|u| + |v|)^2 = |u|^2 + 2|uv| + |v|^2 = 2|z_1|^2 + 2|z_2|^2 + 2|z_1^2 - z_2^2|.$$

Again, by the Parallelogram Identity,

$$2|z_1|^2 + 2|z_2|^2 = |z_1 + z_2|^2 + |z_1 - z_2|^2.$$

Noting that $|z_1^2 - z_2^2| = |z_1 + z_2| \cdot |z_1 - z_2|$, we see that

$$(|u| + |v|)^2 = (|z_1 + z_2| + |z_1 - z_2|)^2.$$

Since both bases are nonnegative, this yields $|u| + |v| = |z_1 + z_2| + |z_1 - z_2|$ which establishes the required result.

Look under Complex Numbers in the Index for similar problems.

S1995-2

Let ω be a complex cube root of unity, with $\omega \neq 1$. Then ω is a root of $x^3 - 1 = (x - 1)(x^2 + x + 1) = 0$, Since $\omega \neq 1$, we must have $\omega^2 + \omega + 1 = 0$. Let S be the given sum,

$$S_1 = \sum_{k=0}^{n-1} \binom{3n}{3k + 1},$$

$$S_2 = \sum_{k=0}^{n-1} \binom{3n}{3k + 2}.$$

Use $\omega^3 = 1$ and the Binomial Theorem to get

$$(1 + 1)^{3n} = S + S_1 + S_2,$$
$$(1 + \omega)^{3n} = S + \omega S_1 + \omega^2 S_2,$$
$$(1 + \omega^2)^{3n} = S + \omega^2 S_1 + \omega S_2.$$

Use $1+\omega+\omega^2 = 0$, so that $(1+\omega)^{3n} = (-\omega^2)^{3n} = (-1)^{3n}\omega^{6n} = (-1)^n$, since $\omega^3 = l$, and $(1 + \omega^2)^{3n} = (-\omega)^{3n} = (-1)^{3n}\omega^{3n} = (-1)^n$, and add the three results above to get

$$3S = 2^{3n} + 2 \cdot (-1)^n.$$

Hence

$$S = \frac{2^{3n} + 2 \cdot (-1)^n}{3}.$$

Look under Finite Sums in the Index for similar problems.

S1995-3

Let $(\sqrt{5} + 2)^p - 2^{p+1} = N + f$, where N is an integer and $0 < f < 1$. We need to show $20p$ divides N. Let $f' = (\sqrt{5} - 2)^p$. Then, since $0 < \sqrt{5} - 2 < 1$, we have $0 < f' < 1$. Hence $-1 < -f' < 0$, and $-1 < f - f' < 1$. But

$$N + f - f' = (\sqrt{5} + 2)^p - 2^{p+1} - (\sqrt{5} - 2)^p$$

$$= 2(p \cdot 1 \cdot 2 \cdot 5^{(p-1)/2} + p \cdot 3 \cdot 2^3 \cdot 5^{(p-3)/2}$$

$$+ \cdots + p \cdot p - 2 \cdot 2^{p-2} \cdot 5 + 2^p) - 2^{p+1}$$

is an integer, since p is odd. Hence $f - f'$ is an integer, which must be $= 0$, since $-1 < f - f' < 1$. Thus

$$N = 4 \cdot 5(p \cdot 1 \cdot 5^{(p-3)/2} + p \cdot 3 \cdot 2^2 \cdot 5^{(p-5)/2} + \cdots + pp - 22^{p-1}.$$

But p is an odd prime implies p divides pr for every r such that $1 \le r \le p - 1$. This shows that $20p$ divides N.

Look under Number Theory in the Index for similar problems.

S1995-4

Let

$$f(x) = 4(x^2 + x + 1)^3 - 27x^2(x + 1)^2.$$

Then

$$f'(x) = 12(x^2 + x + 1)^2(2x + 1) - 27\big(2x(x + 1) + 2x^2(x + 1)\big).$$

We can easily verify that

$$0 = f(1) = f\left(-\frac{1}{2}\right) = f(-2) = f'(1) = f'\left(-\frac{1}{2}\right) = f'(-2).$$

Since $f(x)$ is a sextic, we conclude that 1, $-\frac{1}{2}$, and -2 are double zeroes of $f(x)$ and that $f(x)$ has no other zeros. As the leading coefficient of $f(x)$ is 4, we now see that

$$f(x) = (x - 1)^2 (2x + 1)^2 (x + 2)^2.$$

This establishes the result.

Look under Polynomials in the Index for similar problems.

S1995-5

Let $N = 1^1 \cdot 2^2 \cdot 3^3 \cdots n^n$. Then there are

$$1 + 2 + 3 + \cdots + n = \frac{n(n + 1)}{2}$$

factors in N, their geometric mean is $N^{2/n(n+1)}$, and their arithmetic mean equals

$$\frac{1 + (2 + 2) + (3 + 3 + 3) + \cdots + (n + n + \cdots + n)}{\frac{n(n+1)}{2}}$$

$$= \frac{2(1 + 2^2 + 3^2 + \cdots + n^2)}{n(n + 1)}$$

$$= \frac{2n(n + 1)(2n + 1)}{6n(n + 1)} = \frac{2n + 1}{3}.$$

But the geometric mean of k unequal positive numbers is always less than their arithmetic mean. Hence, $N^{2/n(n+1)} < 2n + 1/3$, that is,

$$N < \left(\frac{2n + 1}{3} \right)^{n(n+1)/2}.$$

Look under Inequalities or Integration in the Index for similar problems.

S1995-6

Let G be the centroid of the triangle. Then G is

$$b + \frac{1}{3} \left(\frac{a\sqrt{3}}{2} \right) = b + \frac{a}{2\sqrt{3}} = c$$

units above L.

$$\text{The area of the triangle} = \frac{1}{2} a \cdot \frac{a\sqrt{3}}{2} = \frac{a^2 \sqrt{3}}{4} \text{ square units.}$$

By Pappus's Theorem, the required volume

$$V = (2\pi c) \frac{a^2\sqrt{3}}{4}$$

$$= \frac{\pi a^2 \sqrt{3}}{2} \left(b + \frac{a}{2\sqrt{3}} \right)$$

$$= \frac{\pi a^2 \sqrt{3}}{2} + \frac{\pi a^3}{4}$$

or $\qquad = \dfrac{\pi a^2}{4}(2\sqrt{3}b + a)$ cubic units.

Look under Volumes in the Index for similar problems.

S1995-7

Solution 1   Let G be the centroid of the triangle. Then G has coordinates

$$\left(\frac{1}{3} \sum_{i=1}^{3} a\cos\alpha_i, \; \frac{1}{3} \sum_{i=1}^{3} a\sin\alpha_i \right).$$

Clearly, by looking at the coordinates of the vertices, we see that the circumcenter of the triangle is the origin, O. Let H be the orthocenter of the triangle. Since, by geometry, O, G, H are collinear and $OG : OH = 1 : 3$, we see that the coordinates of H are

$$\left(\sum_{i=1}^{3} a\cos\alpha_i, \; \sum_{i=1}^{3} a\sin\alpha_i \right).$$

Solution 2   (Solution 2 does not use the property of the Euler line quoted and used in Solution 1. In fact, it proves that O, G, H are collinear and that $OG : OH = 1 : 3$.) Let A, B, C be respectively $(a\cos\alpha_i, a\sin\alpha_i)$, $i = 1, 2, 3$, and H be the orthocenter. Observe that the origin, O, is the circumcenter of triangle ABC. If L is the middle point of BC, then $OL \perp BC$. But $AH \perp BC$ also. Hence $AH \| OL$, i.e., AH is parallel to the vector

$$\left\langle \frac{a}{2}(\cos\alpha_2 + \cos\alpha_3), \frac{a}{2}(\sin\alpha_2 + \sin\alpha_3) \right\rangle$$

$$\left\| \left\langle \cos\left(\frac{\alpha_2 + \alpha_3}{2}\right)\cos\left(\frac{\alpha_2 - \alpha_3}{2}\right), \sin\left(\frac{\alpha_2 + \alpha_3}{2}\right)\cos\left(\frac{\alpha_2 - \alpha_3}{2}\right) \right\rangle \right.$$

$$\left\| \left\langle \cos\left(\frac{\alpha_2 + \alpha_3}{2}\right), \sin\left(\frac{\alpha_2 + \alpha_3}{2}\right) \right\rangle \right..$$

Hence, the parametric equations of AH are

$$\begin{cases} x = a\cos\alpha_1 + t\cos(\frac{\alpha_2+\alpha_1}{2}) \\ y = a\sin\alpha_1 + t\sin(\frac{\alpha_2+\alpha_3}{2}) \end{cases}, \quad t \in R.$$

Similarly, the parametric equations of BH are

$$\begin{cases} x = a\cos\alpha_2 + u\cos(\frac{\alpha_3+\alpha_1}{2}) \\ y = a\sin\alpha_2 + u\sin(\frac{\alpha_3+\alpha_3}{2}) \end{cases}, \quad t \in R.$$

These two lines meet at $H(x,y)$, where

$$t\cos\left(\frac{\alpha_2+\alpha_1}{2}\right) - u\cos\left(\frac{\alpha_3+\alpha_1}{2}\right)$$

$$= -a\cos\alpha_1 + a\cos\alpha_2$$

$$= 2a\sin\left(\frac{\alpha_1-\alpha_2}{2}\right)\sin\left(\frac{\alpha_1+\alpha_2}{2}\right)$$

and

$$t\sin\left(\frac{\alpha_2+\alpha_3}{2}\right) - u\sin\left(\frac{\alpha_3+\alpha_1}{2}\right)$$

$$= -a\sin\alpha_1 + a\sin\alpha_2$$

$$= -2a\sin\left(\frac{\alpha_1-\alpha_2}{2}\right)\cos\left(\frac{\alpha_1+\alpha_2}{2}\right),$$

Solving for t we find,

$$t = 2a\cos\frac{\alpha_3-\alpha_2}{2}.$$

Thus,

$$x = a\cos\alpha_1 + 2a\cos\left(\frac{\alpha_3-\alpha_2}{2}\right)\cos\left(\frac{\alpha_3-\alpha_2}{2}\right)\cos\left(\frac{\alpha_3+\alpha_2}{2}\right)$$

$$= a\cos\alpha_1 + a\cos\alpha_2 + a\cos\alpha_3.$$

Since y is obtained from x by replacing cos with sin, we see that $y = a\sin\alpha_1 + a\sin\alpha_2 + a\sin\alpha_3$. Thus the coordinates of H are

$$\left(\sum_{i=1}^{3} a\cos\alpha_i, \sum_{i=1}^{3} a\sin\alpha_i\right).$$

Look under Geometry in the Index for similar problems.

S1995-8

We have

$$\frac{(1+x)^n}{(1-x)^3} = \left(\sum_{k=1}^{n} \binom{n}{k} x^k \right) \left\{ \frac{1}{2} \sum_{k=0}^{\infty} (k+1)(k+2)x^k \right\}.$$

The coefficient of x^{2n} is

$$\frac{1}{2} \left\{ \binom{n}{0}(2n+1)(2n+2) + \binom{n}{1}(2n)(2n+1) \right.$$

$$\left. + \cdots + \binom{n}{n}(n+1)(n+2) \right\}$$

$$= \frac{1}{2} \sum_{k=0}^{n} \binom{n}{k}(2n-k+1)(2n-k+2)$$

$$= \frac{1}{2} \sum_{k=0}^{n} \binom{n}{k} \{(2n+1)-k\}\{(2n+1)-(k-1)\}$$

$$= \frac{1}{2} \sum_{k=0}^{n} \binom{n}{k}(2n+1)^2 - \frac{2n+1}{2} \sum_{k=0}^{n} \binom{n}{k}(2k-1)$$

$$+ \frac{1}{2} \sum_{k=0}^{n} \binom{n}{k} k(k-1)$$

$$= \frac{1}{2}(4n^2 + 4n + 1 + 2n + 1) \sum_{k=0}^{n} \binom{n}{k} - (2n+1) \sum_{k=0}^{n} \binom{n}{k} k$$

$$+ \frac{1}{2} \sum_{k=0}^{n} \binom{n}{k} k(k-1)$$

$$= \frac{1}{2}(4n^2 + 6n + 2)2^n - (2n+1)(n \cdot 2^{n-1}) + \frac{1}{2}n(n-1)2^{n-2}$$

$$= 2^{n-3} \{4(4n^2 + 6n + 2) - 4n(2n+1) + n^2 - n\}$$

$$= 2^{n-3}(9n^2 + 19n + 8).$$

Note: We have used the following results, obtained by putting $x = 1$ in the binomial expansion of $(1 + x)^n$ and in the results obtained by differentiating with respect to x once and twice.

$$2^n = \sum_{k=0}^{n} \binom{n}{k}, \quad n \cdot 2^{n-1} = \sum_{k=0}^{n} k \cdot \binom{n}{k}, \quad n(n-1) \cdot 2^{n-2} = \sum_{k=0}^{n} k(k-1)\binom{n}{k}.$$

Of course, we have also assumed that $|x| < 1$, so that

$$\frac{1}{1-x} = \sum_{k=0}^{\infty} x^k, \qquad \frac{1}{(1-x)^2} = \sum_{k=0}^{\infty} kx^{k-1},$$

and

$$\frac{1}{(1-x)^3} = \frac{1}{2} \sum_{k=0}^{\infty} k(k-1)x^{k-2}.$$

Exam #31–1996

S1996-1

The problem is equivalent to finding digits (integers from 0 to 9) a, b, c such that $n = 10a + b \geq \sqrt{1000}$ and $n^2 - c^2$ is divisible by 100. Factoring, we see that $(10a + b + c)(10a + b - c)$ is divisible by 100. If 5 divides both factors, then 5 divides their sum $20a + 2b$ and therefore divides b. Otherwise, one of the factors must be divisible by 25, and in fact by 50 since the two factors differ by the even number $2c$. Thus one of $a + (b + c)/10$, $a + (b - c)/10$ is a multiple of 5. Since b and c are digits, $(b \pm c)/10$ must be 0 or 1 and $a = 0, 4, 5, 9$. But for this problem, $10a + b \geq 32$, so either $a = 4, 5, 9$ or $b = 0, 5$. There are $30 + 13 - 6 = 37$ numbers of this form and thus 36 others besides $43^2 = 1849$.

Look under Number Theory in the Index for similar problems.

S1996-2

Let $ABCDE$ denote the pentagon P with $\angle A = 90°$. Then the area of triangle EAB is $1/2$ and quadrilateral $T = EBCD$ has sides $\sqrt{2}, 1, 1, 1, 1$. We may assume that P is convex. Note that

$$\text{area}(EBCD) = \text{area}(BCD) + \text{area}(DEB) = 1/2 \sin C + \sqrt{2}/2 \sin E.$$

Also, by the law of cosines,

$$BD^2 = 2 - 2\cos C = 3 - 2\sqrt{2}\cos E$$

so

$$\frac{1}{2}\cos C - \frac{\sqrt{2}}{2}\cos E = \frac{-1}{4}.$$

Squaring and adding both equations,

$$\left(\text{area}(EBCD)\right)^2 = \frac{11}{16} - \frac{\sqrt{2}}{2}\cos(C + E).$$

This is clearly maximized when $C + E = \pi$, (which is true if and only if $EBCD$ is a trapezoid) and the area of $EBCD$ is

$$\sqrt{\frac{11}{16} + \frac{\sqrt{2}}{2}}.$$

Look under Geometry in the Index for similar problems.

S1996-3

(a) Number the cats from 1 to 100, and consider the nth cat. For each even divisor of n, he will get a dollar and for each odd divisor, he'll lose a dollar. If $n = 2^a \prod_{i=1}^{r} p_i^{a_i}$, then the number of even divisors is $a \prod_{i=2}^{r} (a_i + 1)$ and the number of odd divisors is $\prod_{i=2}^{r} (a_i + 1)$. Hence the nth cat will get $f(n) = (a - 1) \prod_{i=2}^{r} (a_i + 1)$. The cat loses money if $a = 0$, i.e., n is odd, so 50 cats lose money. If $a = 1$, the cat breaks even, and there are 25 of these. Finally, the remaining 25 cats profit.

(b) To maximize $f(n)$, note that if $a < 2$, $f(n) \leq 0$ and if $2^a b \leq 100$, for $a = 6, 5, 4, 3, 2$, then $b \leq 1$, $b \leq 3$, $b \leq 6$, $b \leq 12$, $b \leq 25$. So calculating $f(2^6) = 5$, $f(2^5 \cdot 3) = 8$, $f(2^3 \cdot 3^2) = 6$, and $f(2^2 \cdot 3 \cdot 5) = 4$ gives the 96th cat to gain the most with \$8.

(c) The total amount the rats left with is

$$\sum_{k=1}^{100} (-1)^{k-1} \lfloor \frac{100}{k} \rfloor = 68.$$

Look under Number Theory in the Index for similar problems.

S1996-4

Let the two x coordinates of the square's vertices be a and b with $a > b$. Then $a^3 - 3a = b^3 - 3b = a - b$ so $a = b^3 - 2b$ and $b = 4a - a^3$. Equivalently, $a^3 = 4a - b$ and $b^3 = a + 2b$. Subtracting, $(a - b)(a^2 + ab + b^2) = 3(a - b)$ so $a^2 + ab + b^2 = 3$. Substituting $a = b^3 - 2b$,

$$b^6 - 4b^4 + 4b^2 + b^4 - 2b^2 + b^2 = 3,$$

$$b^6 - 3b^4 + 3b^2 = 3,$$

$$b^6 - 3b^4 + 3b^2 - 1 = 2,$$

$$(b^2 - 1)^3 = 2,$$

$$b^2 = 1 + \sqrt[3]{2}.$$

The area of the square is

$$
\begin{aligned}
(b^3 - 3b)^2 &= b^2(b^2 - 3)^2 \\
&= (1 + \sqrt[3]{2})(\sqrt[3]{2} - 2)^2 \\
&= (1 + \sqrt[3]{2})(\sqrt[3]{4} + 4 - 4\sqrt[3]{2}) \\
&= 6 - 3\sqrt[3]{4} = 6 + \sqrt[3]{-108} \\
&= \sqrt[3]{216} + \sqrt[3]{-108}
\end{aligned}
$$

so $A + B = 108$.

Look under Analytic Geometry in the Index for similar problems.

S1996-5

Reflecting ABC about the y-axis and then the result about the x-axis, gives a closed curve Γ bisecting the area of a square of side length $\sqrt{8}$. Thus Γ encloses a region of area $\frac{1}{2}(\sqrt{8})^2 = 4$. By the isoperimetric theorem, if a closed curve Γ encloses a fixed area, the length of Γ is minimized when Γ is a circle. Therefore Γ is a circle, say of radius r, with $\pi r^2 = 4$, so $r = \sqrt{4/\pi}$ and the length of γ is

$$
\gamma = \frac{1}{4}(2\pi r) = \frac{1}{4}(2\pi)\sqrt{\frac{4}{\pi}} = \sqrt{\pi}.
$$

Look under Geometry in the Index for similar problems.

S1996-6

Such an event is mathematically possible if out of the $2n$ children, exactly n are males. Thus, $p_n = \left(\frac{1}{2}\right)^{2n}\binom{2n}{n}$. The expression $n \cdot p_n^2$ can be written as

$$
\begin{aligned}
np_n^2 &= n(1/2)^{4n}\binom{2n}{n}^2 = n\frac{(2n)!(2n)!}{(2^n n!)^4} \\
&= n\frac{1 \cdot 2 \cdot 3 \cdots 2n}{(2 \cdot 4 \cdot 6 \cdots 2n)^2}\frac{1 \cdot 2 \cdot 3 \cdots 2n}{(2 \cdot 4 \cdot 6 \cdots 2n)^2} \\
&= n\frac{1 \cdot 3 \cdot 5 \cdots (2n - l)}{2 \cdot 4 \cdot 6 \cdots 2n}\frac{1 \cdot 3 \cdot 5 \cdots (2n - l)}{2 \cdot 4 \cdot 6 \cdots 2n} \\
&= n\frac{1 \cdot 3 \cdot 3 \cdot 5 \cdot 5 \cdots (2n - l) \cdots (2n - l)}{2 \cdot 2 \cdot 4 \cdot 4 \cdot 6 \cdots 2n \cdot 2n} \\
&= \frac{1}{2}\frac{3}{2}\frac{3}{4}\frac{5}{4}\frac{5}{6} \cdots \frac{2n-1}{2n-2}\frac{2n-1}{2n}\frac{n}{2n}.
\end{aligned}
$$

Thus,

$$2 \lim_{n \to \infty} n p_n^2 = \frac{1}{2} \frac{3}{2} \frac{3}{4} \frac{5}{4} \frac{5}{6} \cdots = \frac{2}{\pi}$$

by Wallis' formula, and the answer is $1/\sqrt{\pi}$.

An alternate solution can be found using Stirling's Formula: Since $n! \approx \left(\frac{n}{e}\right)^n \sqrt{2\pi n}$,

$$\binom{2n}{n} \approx \frac{\left(\frac{2n}{e}\right)^{2n} \sqrt{2\pi 2n}}{\left[\left(\frac{n}{e}\right)^n \sqrt{2\pi n}\right]^2} = \frac{2^{2n}}{\sqrt{\pi n}}$$

whence $\sqrt{n} \cdot p_n \approx \frac{1}{\sqrt{\pi}}$.

Look under Limit Evaluation or Probability in the Index for similar problems.

S1996-7

If $\{x_1, x_2, x_3, x_4\}$ is statistically equivalent to $\{1, 9, 9, 6\}$ then

$$\frac{1}{4} \sum_{i=1}^{4} x_i = \frac{1}{4}(1 + 9 + 9 + 6)$$

so that $\sum_{i=1}^{4} x_i = 25$. If σ is the standard deviation, then

$$\sum_{i=1}^{4} x_i^2 = 3\sigma^2 + \frac{1}{4} \left(\sum_{i=1}^{4} x_i \right)^2$$

and we must have

$$\sum_{i=1}^{4} x_i^2 = 1^2 + 9^2 + 9^2 + 6^2 = 199.$$

Letting $y_i = x_i - 6$,

$$\sum_{i=1}^{4} y_i = 1 \quad \text{and} \quad \sum_{i=1}^{4} y_i^2 = 43.$$

Since squares are congruent to either 0 or 1 mod 4, three of the y_i (say y_1, y_2, y_3) must be odd and one (say y_4) must be even. As odd squares are congruent to 1 mod 8, we have $y_4^2 + 3 \equiv 43 \mod 8$ whence y_4 must be a multiple of 4. Since each $|y_i| < 7$, $|y_4| = 0$ or 4. If $|y_4| = 4$, then we must have three odd squares which add to 27. This is possible in two different ways: $27 = 25 + 1 + 1 = 9 + 9 + 9$. Checking to see if any of the signed sums of the elements from $\{5, 4, 1, 1\}$ or $\{4, 3, 3, 3\}$ equals 1, we find $5 - 4 + 1 - 1 = 1$, $4 - 5 + 1 + 1 = 1$, and $4 - 3 + 3 - 3 = 1$. The $x_i = y_i + 6 = \{11, 2, 7, 5\}$, $\{10, 1, 7, 7\}$, and $\{10, 3, 9, 3\}$. If $|y_4| = 0$, we must have three odd squares which add to 43. This is possible only one way: $43 = 9 + 9 + 25$. Checking to see if any of the signed sums of the elements from $\{0, 3, 3, 5\}$ equals 1, we find $0 + 3 + 3 - 5 = 1$. Then $x_i = y_i + 6 = \{1, 9, 9, 6\}$.

Look under Diophantine Equations in the Index for similar problems.

SI996-8

Let $(a, f(a))$ and $(b, f(b))$ be arbitrary points on the graph with $b > a$ and let $c = \frac{a+b}{2}$. We are given that

$$\frac{f(a) + f(b)}{2} - f(c) = (b - a)^2,$$

or equivalently,

$$\frac{\dfrac{f(b) - f(c)}{b - c} - \dfrac{f(c) - f(a)}{c - a}}{b - a} = 8.$$

Take the limit as $b \to a$ (and a fortiori $b \to c$, $c \to a$) to see that $f''(a) = 8$. Since a was arbitrary, we have $y = 4x^2 + Ax + B$.

Look under Differentiation in the Index for similar problems.

Exam #32–1997

SI997-I

We will consider combinations of the last four digits only. If the four digits are all different, then there are $\binom{10}{4} = 210$ combinations. If we have three different digits, then there are $3 \cdot \binom{10}{3} = 360$ combinations (e.g., the digits $\{1, 2, 3\}$ generate three combinations, 1123, 1223, and 1233). If we have two different digits, then there are $3\binom{10}{2} = 135$ combinations (e.g., the digits $\{1, 2\}$ generate three combinations 1112, 1122, and 1222). Finally, if all the digits coincide then there are ten combinations. The total number of combinations is $210 + 360 + 135 + 10 = 715$ and the probability is $p = 715/10{,}000 = 0.0715$

Look under Enumeration or Probability in the Index for similar problems.

S1997-2

We calculate sums directly,

$$\sum_{i-1}^{n}\sum_{j=1}^{i} j = \sum_{i=1}^{n} \frac{i(i+1)}{2} = \frac{1}{2}\sum_{i=1}^{n} i(i+1) = \frac{1}{2}\left(\sum_{i=1}^{n} i^2 + \sum_{i=1}^{n} i\right).$$

By using two well-known formulas the above is equal to

$$\frac{1}{2}\left(\frac{n(n+1)(2n+1)}{6} + \frac{n(n+1)}{2}\right).$$

Simplifying we get our answer $n(n+1)(n+2)/6$. It is interesting to note that these numbers appear as the first n numbers in the 3rd diagonal of Pascal's triangle.

Look under Finite Sums in the Index for similar problems.

S1997-3

Note first that $z^4 - 5z^2 + 6 = (z^2 - 3)(z^2 - 2)$. So

$$|z^4 - 5z^2 + 6| = |(z^2 - 3)(z^2 - 2)| = |z^2 - 3||z^2 - 2|.$$

But we know that $|z_1 - z_2| \geq ||z_1| - |z_2||$ for all z_1 and z_2 in C. So

$$|z^2 - 3| \geq ||z^2| - 3| = ||z|^2 - 3| = |9 - 3| = 6$$

and

$$|z^2 - 2| \geq ||z^2| - 2| = ||z|^2 - 2| = |9 - 2| = 7.$$

Thus, $|z^2 - 3||z^2 - 2| \geq (6)(7) = 42$. Finally,

$$\frac{21}{|z^4 - 5z^2 + 6|} \leq \frac{21}{42} = \frac{1}{2}.$$

So letting $A = 1/2$ we get an upper bound. Further, A is the smallest upper bound since if $z = \pm 3$, then equality holds.

Look under Complex Numbers in the Index for similar problems.

S1997-4

If $x^y = y^x$ for $x, y > 0$, then taking the natural logarithm of each side we obtain $y \ln x = x \ln y$, which implies

$$\frac{\ln x}{x} = \frac{\ln y}{y}.$$

Obviously all the above is true if $x = y$. So suppose $x \neq y$. Then if we graph the function $v = \ln u$, we see that the line through the points $(x, \ln x)$ and $(y, \ln y)$ goes through the origin and has equation

$$v = mu.$$

Hence $\ln y = my$ and $\ln x = mx$. We also note that this is possible only if $x, y > 1$ and $m > 0$. Let $y = tx$ where $t > 0$ and $t \neq 1$. Then

$$\ln tx = mtx,$$
$$\ln t + \ln x = t(mx),$$
$$\ln t + \ln x = t \ln x.$$

Simplifying we get
$$t^{1/(t-1)} = x.$$

Then $y = tx = t^{t/(t-1)}$. So all pairs of the form $(t^{1/(t-1)}, t^{t/(t-1)})$ satisfy the original equation. (If we restrict $t > 1$ we get all solutions with $x < y$.) Some pairs include:

$$t = 2 \Rightarrow (2, 4),$$
$$t = 3 \Rightarrow (\sqrt{3}, 3\sqrt{3}),$$
$$t = 3/2 \Rightarrow (9/4, 27/8).$$

S1997-5

We claim that 16 points and 16 lines satisfy the six axioms. The proof that this is the fewest number of points and lines will be deferred to the solution of part (b) below. The six axioms will guide us in constructing what is essentially a 4×4 array. Each row (and column) will contain four points, thus giving us 16 points in our solution. The fact that we also have 16 lines will become apparent later. Let us begin the construction.

By Axiom 5, there is at least one line, call it r_1. By Axiom 1, there are exactly four points on this line, call them A_1, B_1, C_1, and D_1. We adopt the convention that a line in our system can be described by the four points that it contains. So in particular, we write $r_1 = A_1 B_1 C_1 D_1$. Now by Axiom 3 and Axiom 2, there is a line c_1 that contains the point A_1 but does not contain any of the points B_1, C_1, and D_1. Let A_2, A_3, and A_4 be the other three points on line c_1. By Axiom 6 and Axiom 2, there are three new lines $r_2 = A_2 B_2 C_2 D_2$, $r_3 = A_3 B_3 C_3 D_3$, and $r_4 = A_4 B_4 C_4 D_4$. These lines together add a total of nine new points, namely, B_i, C_i, and D_i for $1 \leq i \leq 3$. We now have the 16 points in our system. By Axiom 6, we now add three new lines, $c_2 = B_1 B_2 B_3 B_4$, $c_3 = C_1 C_2 C_3 C_4$, and

$c_4 = D_1 D_2 D_3 D_4$. It is now easy to picture the 4×4 array: we have four row lines r_1, r_2, r_3, and r_4 and four column lines c_1, c_2, c_3, and c_4. Note that this simple construction satisfies all of the axioms except Axiom 3. To satisfy Axiom 3, we must add more lines (this is the tricky part).

We start by satisfying Axiom 3 for the point A_1. The possible lines through A_1 are:

   (a) $A_1 B_2 C_3 D_4$,   (b) $A_1 B_2 D_3 C_4$,   (c) $A_1 C_2 B_3 D_4$,

   (d) $A_1 C_2 D_3 B_4$,   (e) $A_1 D_2 C_3 B_4$,   (f) $A_1 D_2 B_3 C_4$.

Let us choose line (a). Then by Axiom 2, we must exclude lines (b), (c), and (e). However, we can choose line (d) and that gives us four lines that contain point A_1. We now want to satisfy Axiom 3 for the point B_1. As above, Axiom 2 limits the possible lines that we can choose. The possible lines through B_1 are:

   (1) $B_1 A_2 D_3 C_4$,   (2) $B_1 C_2 A_3 D_4$,   (3) $B_1 D_2 C_3 A_4$.

We choose lines (1) and (3) to satisfy Axiom 3 for B_1. Note that Axiom 3 is also satisfied for the points C_3 and D_3. In a similar fashion we choose the following four lines to complete our system:

   (i) $C_1 A_2 B_3 D_4$,   (ii) $C_1 D_2 A_3 B_4$,

   (iii) $D_1 B_2 A_3 C_4$,   (iv) $D_1 C_2 B_3 A_4$.

This system of 16 points and 16 lines satisfies all of the axioms.

For each positive integer n, the problem statement associates a set of six axioms to n. Say that a collection of points and lines satisfying these axioms is an n-configuration. If C is an n-configuration and x, y are lines in C, we say that x and y are parallel if they share no points in C. For each line, the set consisting of the line and all those parallel to it is a family of parallels. For $n = 1$, there is a configuration with two points and two lines that satisfies the axioms, and no configuration with fewer points or fewer lines will do. The rest of the solution to part (b) follows from the following:

Theorem.   *Let n be a positive integer with $n > 1$. If an n-configuration exists, then there is one with n^2 points and n^2 lines, and no n-configuration with fewer points or fewer lines exists.*

Proof.   Let n be an integer, $n > 1$, and let C be an n-configuration. Let x be any line of C and choose a point P on x. By Axiom 3, there exists another line y that contains p. By Axiom 1, y contains exactly n points, say $p_1 = P, p_2, \ldots, p_n$. For each $i, 2 \leq i \leq n$, there is by Axiom 6 a unique line x_i containing p_i and parallel to x. Let x_1 denote x.

Using Axiom 6, it is easy to verify that "equal to or parallel to" defines an equivalence relation on the set of lines of C, so x_i is parallel to x_j when $i \neq j$. Thus, the n lines $x_1, \ldots, x_n$ together contain n^2 points. We wish to show that these are all the points of C. To this end, let Q be any point of C that is not on x_1. Then there exists (by Axiom 6) a unique line z through Q and parallel to x_1. Again by Axiom 6, there cannot be two lines through P and parallel to z. So since z is parallel to x_1, necessarily z meets y, say at p_j. But then z must be x_j, so Q is on x_j, which is a contradiction.

Therefore C has exactly n^2 points. It is then clear that every line has exactly n lines in its family of parallels.

Let k be the number of families of parallels for C. Since each point is on exactly one line from each family, Axiom 3 implies $k \geq n$. This shows that C must have at least n^2 lines. Suppose that $k > n$. Choose any family of parallel lines and discard it to obtain a collection B of n^2 points and $k - 1$ families of parallel lines. Since we discarded an entire family of parallels, B satisfies Axiom 6. Each point of B is on at least n lines of B, so Axiom 3 is satisfied. Since $n > 1$ and C is an n-configuration, B satisfies Axioms 1, 2, 4, and 5, and thus B is an n-configuration. It is clear that we may continue in this fashion and discard families of parallel lines until we have a minimum of n families of parallel lines. The resulting n-configuration then has n^2 points and n^2 lines.

We have shown that any n-configuration must have at least (exactly) n^2 points and at least n^2 lines. We also showed that if an n-configuration exists, then there is one with exactly n^2 points and exactly n^2 lines. Thus n^2 is the fewest number of points and lines for an n-configuration. This proves the Theorem.

Look under Synthetic Geometry in the Index for similar problems.

S1997-6

Part **(a)** is a standard exercise in norms. The first set of points $\{x \in R^2 : \|x\|_1 = 1\}$ is the square in the plane with vertices $(1, 0)$, $(0, 1)$, $(-1, 0)$, and $(0, -1)$. The second set of points $\{x \in R^2 : \|x\|_\infty = 1\}$ is the square in the plane with vertices $(1, 1)$, $(-1, 1)$, $(-1, -1)$, and $(1, -1)$.

For part **(b)**, we first consider the 1-norm. If $\overline{x} = (x_1, x_2)$ and $\overline{y} = (y_1, y_2)$, then $\overline{x} + \overline{y} = (x_1 + y_1, x_2 + y_2)$. Then

$$\|\overline{x} + \overline{y}\|_1 = \|\overline{x}\|_1 + \|\overline{y}\|_1$$

holds if and only if

$$|x_1 + y_1| + |x_2 + y_2| = |x_1| + |x_2| + |y_1| + |y_2|.$$

Equality holds in this second equation if and only if

$$|x_1 + y_1| = |x_1| + |y_1| \quad \text{and} \quad |x_2 + y_2| = |x_2| + |y_2|.$$

Thus, x_1 and y_1 must have the same sign and similarly for x_2 and y_2. Geometrically, this means that $\overline{x}$ and $\overline{y}$ must be in the same quadrant.

For the ∞-norm,

$$\|\overline{x} + \overline{y}\|_\infty = \|\overline{x}\|_\infty + \|\overline{y}\|_\infty$$

holds if and only if

$$\max\{|x_1 + y_1|, |x_2 + y_2|\} = \max\{|x_1|, |x_2|\} + \max\{|y_1|, |y_2|\}.$$

Let $|x_i + y_i| = \max\{|x_1 + y_1|, |x_2 + y_2|\}$. Then

$$|x_i + y_i| \le |x_i| + |y_i| \le \max\{|x_1|, x_2|\} + \max\{|y_1|, |y_2|\}$$

and equality holds if and only if each of the following holds:

(1) $|x_i| = \max\{|x_1|, |x_2|\}$,

(2) $|y_i| = \max\{|y_1|, |y_2|\}$,

(3) x_i and y_i have the same sign.

The solution is easier to view geometrically. The diagonals $y = x$ and $y = -x$ partition the plane into the following four parts:

(a) $R_1 = \{(x_1, x_2) \in R^2 : x_2 \ge |x_1|\}$,

(b) $R_2 = \{(x_1, x_2) \in R^2 : x_2 \le -|x_1|\}$,

(c) $R_3 = \{(x_1, x_2) \in R^2 : x_1 \ge |x_2|\}$,

(d) $R_4 = \{(x_1, x_2) \in R^2 : x_1 \ge -|x_2|\}$.

Then

$$\|\overline{x} + \overline{y}\|_\infty = \|\overline{x}\|_\infty + \|\overline{y}\|_\infty$$

holds if and only if x and y lie in the same part. The proof easily follows from the properties (1), (2), and (3) above.

Look under Analytic Geometry in the Index for similar problems.

S1997-7

The volume of water can be found by multiplying the cross-sectional area of the remaining water with the length (4 feet) of the trough. Thus, it

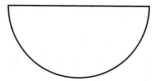

suffices to find the cross-sectional area of the remaining water when the trough is tilted through an angle α with the horizontal, as illustrated.

To find the cross-sectional area of the remaining water (region W) we subtract the areas of the sector on the left and the isosceles triangle in the middle of the figure above. One can show that the angle subtended by the arc in the sector is 2α. So the area of the sector is

$$A_{sector} = \frac{1}{2}(2\alpha)r^2 = \alpha.$$

The isosceles triangle has congruent angles of measure α. Using trigonometry, we get the area of the triangle

$$A_{triangle} = \frac{1}{2}bh = \frac{1}{2}(2\cos\alpha)(\sin\alpha) = \cos\alpha\sin\alpha.$$

We subtract these areas from the total area of the cross-section, which is $\pi/2$, to get $\pi/2 - \cos\alpha\sin\alpha$ for the cross-sectional area of the remaining water. Now multiplying by the length of the trough, we get the volume of the remaining water $2\pi - 4\alpha - 4\cos\alpha\sin\alpha$ feet^3.

Look under Geometry in the Index for similar problems.

Exam #33–1998

S1998-1

Here is a Cartesian proof: put P at the origin with AB along the x-axis and C, D, E in the upper half-plane. Let C have equation $(x-a)^2 + y^2 = r^2$, and let the lines PE and PC have equations $y = bx$ and $y = -bx$, respectively. Note that the slopes of these two lines are opposite since D must lie on the y-axis and $\angle CPD = \angle DPE$. The coordinates of E and C, then, are (x_1, bx_1) and $(x_2, -bx_2)$, respectively, where x_1 and x_2 are the positive and negative (resp.) roots of the equation $(x-a)^2 + b^2x^2 = r^2$. Furthermore, $D = (0, \sqrt{r^2 - a^2})$. Now, we calculate distances: $PE = x_1\sqrt{1+b^2}$, $PC = -x_2\sqrt{1+b^2}$. Since x_1 and x_2 are solutions of the same quadratic equation, we know that their product is the ratio of the constant term of that equation to the leading term. Therefore,

$$PE \cdot PC = -x_1 x_2 (1 + b^2) = \frac{-(a^2 - r^2)(1 + b^2)}{(1 + b^2)} = r^2 - a^2 = PD^2.$$

A transformational proof: reflect through AB to obtain C'', D'' and E'', then observe that PE'' is a continuation of PC and PC'' is a continuation of PE. Apply "mean proportional" to the chords DD'' and CE''.

Look under Geometry in the Index for similar problems.

S1998-2

Let $k = n/7$ and $m = (\frac{n-u}{10} - 2u)/7$. Then, $70m = 7k - 21u$ so that k is an integer if and only if m is an integer .

Look under Number Theory in the Index for similar problems.

S1998-3

One direction is obvious: if $n = p^2 - q^2$, then $n = (p + q)(p - q)$ and both of these factors are greater than 1. In the other direction, if $n = st$ with $s \geq t > 1$, then we set $p = (s + t)/2$ and $q = (s - t)/2$. Since n is odd, both s and t must be odd, so that both their sum and difference are even. Thus, p and q are nonnegative integers, with $n = p^2 - q^2$. To see that $p - q > 1$ observe that $p - q = t$.

Look under Number Theory in the Index for similar problems.

S1998-4

For part **(a)**, use induction. First, observe that $(0, 1)$ is on one of the hyperbolas. Then, assume that (f_{k-2}, f_{k-1}) is on one of the hyperbolas, that is, $f_{k-1}^2 - f_{k-1} f_{k-2} - f_{k-2}^2 = \pm 1$.

Then, we verify that

$$f_k^2 - f_k f_{k-1} - f_{k-1}^2$$
$$= (f_{k-1} + f_{k-2})^2 - (f_{k-1} + f_{k-2}) f_{k-1} - f_{k-1}^2$$
$$= f_{k-1}^2 + 2 f_{k-1} f_{k-2} + f_{k-2}^2 - f_{k-1}^2 - f_{k-1} f_{k-2} - f_{k-1}^2$$
$$= -f_{k-1}^2 + f_{k-1} f_{k-2} + f_{k-2}^2$$
$$= \mp 1.$$

Thus, inductively, we see that all points of $\Im$ are on one of these hyperbolas (and they alternate between the two).

For part **(b)**, we will repeatedly need the following solution of y in terms of x for first quadrant points (the choice of sign corresponds to the choice of hyperbola):

$$y = \frac{x + \sqrt{5x^2 \pm 4}}{2} = x\frac{1 + \sqrt{5 \pm 4/x^2}}{2}.$$

Now, using this, we observe a few preliminary facts: if (a, b) is a first-quadrant integer point on one of these hyperbolas and $a \geq 2$, then $b \geq a$ (since $5 \pm 4/a^2 \geq 4$). Furthermore, if $a \geq 2$, then $b \leq 2a$ (since $5 \pm 4/a2 \leq 9$). Also, if $a < 2$ and is a nonnegative integer, then we have only a few possibilities to consider ($a = 0$ and $a = 1$) with both sign choices), leading to three integer points: $(0, 1), (1, 1)$ and $(1, 2)$, all of which are in **F**.

Next, assume that there is a point (a, b) which is an integer, first-quadrant point on one of the two hyperbolas, but is not in **F**. Let (a_0, b_0) be such a point with a_0 as small as possible. From above, we know that $a_0 \geq 2$ and that $a_0 \leq b_0 \leq 2a_0$. Consider the point $(b_0 - a_0, a_0)$: it is clear that this is still a first-quadrant integer point and that $b_0 - a_0 \leq a_0$. We now show that this point is on one of the hyperbolas:

$$a_0^2 - (b_0 - a_0)a_0 - (b_0 - a_0)^2 = a_0^2 - b_0 a_0 + a_0^2 - b_0^2 + 2b_0 a_0 - a_0^2$$
$$= a_0^2 + b_0 a_0 - b_0^2 = \pm 1$$

since (a_0, b_0) is on one of the hyperbolas. Furthermore, there is exactly one point at which $b_0 = 2a_0$, namely,(1,2) so that $(b_0 - a_0, a_0)$ has a strictly smaller first coordinate than (a_0, b_0), implying by our assumption that it must be in **F**. But, then (a_0, b_0) must have been in **F** as well, contradicting our choice of (a_0, b_0).

Look under Analytic Geometry or Fibonnacci Sequences in the Index for similar problems.

S1998-5

Denote by $A_1(x)$ the area of the portion of $R(x)$ lying between f_0 and f_1. Denote by $A_2(y)$ the area of the portion of $R(f_1^{-1}(y))$ lying between f_1 and f_2. Then,

$$A_1(x) = \int_0^x f_1(t) - f_0(t)\, dt = \int_0^x (\alpha - 1)t^\beta\, dt = \frac{(\alpha - 1)}{(\beta - 1)}x^{\beta+1}.$$

Furthermore,

$$A_2(y) = \int_0^y f_1^{-1}(t) - f_2^{-1}(t)\, dt = \int_0^y \frac{t^{1/\beta}}{\alpha^{1/\beta}} - f_2^{-1}\, dt.$$

Now, the condition that f_1 bisect f_0 and f_2 in area may be rephrased as $A_2(y) = A_1(f_1^{-1}(y))$ or

$$\int_0^y \frac{t^{1/\beta}}{\alpha^{1/\beta}} - f_2^{-1}\, dt = \frac{(\alpha-1)y^{\frac{\beta+1}{\beta}}}{(\beta+1)\alpha^{\frac{\beta+1}{\beta}}}.$$

Integrating and solving a bit further, we find that

$$\int_0^y f_2^{-1}(t)\, dt = y^{\frac{\beta+1}{\beta}}\frac{\alpha\beta+1-\alpha}{(\beta+1)\alpha^{\frac{\beta+1}{\beta}}}.$$

Solving for f_2^{-1} and then inverting, we find that

$$f_2(x) = \frac{x^\beta \alpha^{\beta+1}\beta^\beta}{(\alpha\beta+1-\alpha)^\beta}.$$

Look under Integration in the Index for similar problems.

S1998-6

(a)   An inclusion-exclusion argument gives

$$D_n = n! - n(n-1)! + \binom{n}{2}(n-2)! - \binom{n}{3}(n-3)! + \cdots.$$

Rewriting this a bit gives

$$D_n = n! \sum_{i=0}^n \frac{(-1)^i}{i!}.$$

This latter form is useful for part **(b)**: Since

$$e = \sum_{i=0}^\infty \frac{(-1)^i}{i!},$$

it is clearly sufficient for part (b), to show that

$$|n!/e - D_n| = |\sum_{i=n+1}^\infty \frac{(-1)^i n!}{i!}| < 1/2.$$

Since this latter series is an alternating series with terms decreasing in absolute value, its limit has absolute value strictly less than the absolute value of the first term which is $1/(n+1)$. So, for $n > 0$ we have part (b).

For part **(c)**, using the expression above, together with the ratio test method for computing radius of convergence, yields the following expression for the radius:

$$\lim_{n\to\infty} \frac{D_{n+1}n!}{D_n(n+1)!} = \lim_{n\to\infty} \frac{n!(n+1)!\sum_{i=0}^{n+1}\frac{(-1)^i}{i!}}{(n+1)!n!\sum_{i=0}^{n}\frac{(-1)^i}{i!}}$$

$$= \lim_{n\to\infty} 1 + \frac{(-1)^{n+1}}{(n+1)\sum_{i=0}^{n}\frac{(-1)^i}{i!}} = 1.$$

Look under Derangements in the Index for similar problems.

S1998-7

For part **(a)**, one very simple method is to observe that the example generalizes in the following way: if $\det A = 1$, then $(A\pm I)^2 = (\operatorname{tr} A\pm 2)A$ so that one simple method of coming up with two square roots of $[A]$ is to find a representative of the class, say A', which has determinant 1 (do this by dividing A by the square root of its determinant, which is possible since we're dealing with complex matrices), then adding or subtracting I. An alternate form would be

$$\sqrt{[A]} = [A \pm \sqrt{\det A}\, I].$$

Generally, this will produce two distinct valid square roots, but occasionally one of the two will be singular. This occurs when $\mp\sqrt{\det A}$ is an eigenvalue of A which occurs only when A has a repeated eigenvalue. In this case, one of the two square roots will be singular (and its square will be the zero matrix).

(b) This provides us with one or two distinct square roots. Is this all? The answer, for classes other than the class of the identity, is "yes." The way to see this is to take a representative of the equivalence class which has determinant 1 and put it in Jordan canonical form, say $C = BAB^{-1}$ is in Jordan form. Any square roots $[D]$ of $[C]$ correspond to square roots $[B^{-1}DB]$ of $[A]$. So, we are reduced to examining determinant 1 Jordan forms. There are two such, diagonal and nondiagonal.

Consider the nondiagonal form first. Here $C = \begin{pmatrix} 1 & 1 \\ 0 & 1 \end{pmatrix}$ and we are looking for $D = \begin{pmatrix} a & b \\ c & d \end{pmatrix}$ such that $D^2 = \omega C$ for some ω.

We calculate that

$$\begin{pmatrix} a & b \\ c & d \end{pmatrix}^2 = \begin{pmatrix} a^2+bc & b(a+d) \\ c(a+d) & d^2+bc \end{pmatrix}.$$

For this to be equal to $\begin{pmatrix} \omega & \omega \\ 0 & \omega \end{pmatrix}$ we need $c(a+d) = 0$ and $b(a+d) = a^2 + bc = d^2 + bc = \omega \neq 0$. The first equation (together with one of the latter) implies that $c = 0$ and $a + d \neq 0$. But, since clearly $a^2 = d^2$, we must have $a = d$. Thus we are reduced to $2ab = a^2$ which gives us one equivalence class, namely $\left[\begin{pmatrix} 2 & 1 \\ 0 & 2 \end{pmatrix} \right]$.

In the diagonal case, $C = \begin{pmatrix} \lambda & 0 \\ 0 & \lambda^{-1} \end{pmatrix}$ with $\lambda \neq \lambda^{-1}$ (otherwise C is equivalent to I). So, we have $b(a+d) = c(a+d) = 0$, $a^2 + bc = \omega\lambda$ and $d^2 + bc = \omega\lambda^{-1}$. The former implies that either $b = c = 0$ or $a + d = 0$, but if $a = -d$, the latter two cannot hold, since this would force $\omega\lambda = \omega\lambda^{-1}$. So we must have $b = c = 0$, $a^2 = \omega\lambda$ and $d^2 = \omega\lambda^{-1}$. This gives us two classes of square roots,

$$\left[\begin{pmatrix} \sqrt{\lambda} & 0 \\ 0 & 1/\sqrt{\lambda} \end{pmatrix} \right] \quad \text{and} \quad \left[\begin{pmatrix} \sqrt{\lambda} & 0 \\ 0 & -1/\sqrt{\lambda} \end{pmatrix} \right]$$

which are equivalent to the two square roots computed by the simpler method above.

For part **(c)**, we apply the calculations for part (b) and obtain:

$$b(a+d) = c(a+d) = 0, \quad a^2 + bc = d^2 + bc.$$

These are all satisfied whenever $a + d = 0$, giving infinitely many equivalence classes of square roots of $[I]$, in addition to the obvious fact that $[I]$ is its own square root. To show that there are infinitely many nonequivalent square roots, consider the matrices $\begin{pmatrix} 1 & a \\ b & -1 \end{pmatrix}$ where a, b are arbitrary complex numbers.

Look under Matrix Algebra in the Index for similar problems.

Exam #34–1999

S1999-1

The key observation is that the procedure never alters the parity (even/odd count) of the orange balls. Each performance results in a loss of one ball from the urn, so that, eventually, just one is left. Its color must be orange since the original number of such balls is odd.

Look under Logic in the Index for similar problems.

S1999-2

Various correct proofs can be given, some using trigonometry, some via analytic geometry. The simplest and most elegant proof, however, is to consider *area*. We have

area of T = area of ABP + area of BCP + area of CAP,

where A, B, and C are the vertices of T. The standard area formula ($\frac{1}{2}$base × height) for triangles then shows that the constant of the problem is the altitude of T.

Look under Geometry in the Index for similar problems.

S1999-3

The inequality may be rewritten as

$$\sqrt{(2x)^2 + (x+y)^2} \le \sqrt{2}x + \sqrt{x^2 + y^2}. \qquad \text{(i)}$$

We square both sides, in attempting to eradicate the radicals, giving

$$5x^2 + 2xy + y \le 2x^2 + 2\sqrt{2}x\sqrt{x^2 + y^2} + x^2 + y^2. \qquad \text{(ii)}$$

Cancellation leads to the simpler inequality

$$x + y \le \sqrt{2(x^2 + y^2)}. \qquad \text{(iii)}$$

Squaring again gives

$$x^2 + 2xy + y \le 2(x^2 + y^2) \qquad \text{(iv)}$$

or

$$0 \le (x - y)^2. \qquad \text{(v)}$$

Inequalities (i) to (v) are all equivalent. (v) is valid, with equality only when $y = x$, so that the same is true of (i).

Look under Inequalities in the Index for similar problems.

S1999-4

There is a grain of truth to the claim mentioned in the question. *If f is a convex function*, increasing or not, then the Riemann sums (over uniform partitions) do tend monotonically to the integral. This is true whether the sums are evaluated at the right-hand endpoints (as they are here), or at the left, or whether they are inscribed or circumscribed. A counterexample to the claim is provided by the piecewise linear function

$$f(x) = \begin{cases} 0, & \text{if } 0 \le x \le 1/2, \\ 6x - 3, & \text{if } 1/2 \le x \le 2/3, \\ 1, & \text{if } 2/3 \le x \le 1. \end{cases}$$

It is clear that

$$\frac{f(1)}{1} = 1, \qquad \frac{f(1/2) + f(1)}{2} = 1/2,$$

and

$$\frac{f(1/3) + f(2/3) + f(1)}{3} = 2/3.$$

Look under Riemann Sums in the Index for similar problems.

S1999-5

We take the dartboard to be the square centered at the origin with corners at $(\pm 1, \pm 1)$. The points closer to the center than the edge are those enclosed by four parabolic arcs as illustrated.

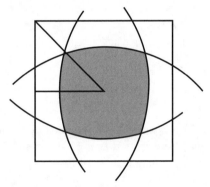

The desired probability is the ratio of the shaded area to the total area of the square. The symmetry of the problem enables us to compute this ratio by restricting our attention to the triangle indicated below.

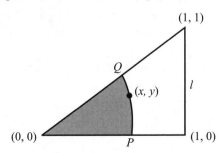

The parabolic arc is the locus of points equidistant from $(0,0)$ and the line l. We have
$$\sqrt{x^2 + y^2} = 1 - x,$$
so that the equation of the arc is $x = (1 - y^2)/2$.

P is the point $(1/2, 0)$, while Q is the intersection of the arc with the line $y = x$. Letting $Q = (q, q)$, we must have $q = (1 - q^2)/2$, so that $Q = (\sqrt{2} - 1, \sqrt{2} - 1)$.

The shaded area of the triangle is
$$\int_0^{\sqrt{2}-1} \left(\frac{1 - y^2}{2} - y \right) dy = \left[\frac{y}{2} - \frac{y^3}{6} - \frac{y^2}{2} \right] \Big|_0^{\sqrt{2}-1} = \frac{4\sqrt{2} - 5}{6}.$$

The total area of the triangle is $1/2$, so the probability is $\frac{4\sqrt{2}-5}{3}$.

Look under Probability in the Index for similar problems.

S1999-6

Three consecutive entries of the nth row, say
$$\binom{n}{k}, \quad \binom{n}{k+1}, \quad \text{and} \quad \binom{n}{k+2},$$
stand in the ratio $1 : 2 : 3$ if and only if
$$\frac{n - k}{k + 1} = 2 \quad \text{and} \quad \frac{n - k - 1}{k + 2} = \frac{3}{2}.$$
These equations, obtained by expanding the binomial coefficients as prescribed in the question, may be rephrased as $n = 3k + 2$ and $2n = 5k + 8$.

There is one, and only one, solution: $n = 14$, $k = 4$.

S1999-7

The series diverges. Its terms lie between those of the (divergent) harmonic series, $\sum_{k=1}^\infty 1/k$, and those of the (convergent) alternating harmonic series, $\sum_{k=1}^\infty (-1)^k/k$. There is some cancellation in our series, courtesy of the cosine, but this occurs too slowly for convergence because of the logarithm.

We show that the partial sums of our series are not Cauchy: more precisely, we prove
$$\sum_{k=e^{(2n-\frac{1}{4})\pi}}^{e^{(2n+\frac{1}{4})\pi}} \frac{\cos(\log k)}{k} > \frac{1 - e^{-\pi/2}}{\sqrt{2}} \quad (n = 1, 2, \ldots). \quad (*)$$

To do this, we observe that

$$\cos x \geq \frac{1}{\sqrt{2}}$$

whenever

$$\left(2n - \frac{1}{4}\right)\pi \leq k \leq \left(2n + \frac{1}{4}\right)\pi$$

so that

$$\cos(\log k) \geq \frac{1}{\sqrt{2}}$$

whenever

$$e^{(2n-\frac{1}{4})\pi} \leq k \leq e^{(2n+\frac{1}{4})\pi}.$$

The series $(*)$ contains at least

$$e^{(2n+\frac{1}{4})\pi} - e^{(2n-\frac{1}{4})\pi}$$

terms, all of which are greater than

$$\frac{1}{\sqrt{2}e^{(2n+\frac{1}{4})\pi}}.$$

Look under Infinite Series in the Index for similar problems.

S1999-8

The standard formula for the sum of a finite geometric series has a matrix analogue,

$$(I + A + \cdots + A^{m-1})(I - A) = I - A^m.$$

$I - A^m = O$ by hypothesis, so that

$$(I + A + \cdots + A^{m-1})(I - A)\mathbf{x} = \mathbf{0} \tag{$**$}$$

no matter what the vector $\mathbf{x}$. Since $I - A \neq 0$, there exist nonzero vectors $\mathbf{x}$ and $\mathbf{y}$ with

$$(I - A)\mathbf{x} = \mathbf{y}.$$

But then, by $(**)$,

$$(I + A + \cdots + A^{m-1})\mathbf{y} = \mathbf{0}$$

so that $I + A + \cdots + A^{m-1}$ is singular and $\det(I + A + \cdots + A^{m-1}) = 0$.

Look under Matrix Algebra in the Index for similar problems.

Exam #35–2000

S2000-1

If Cindy didn't do it, the other three are lying. In particular, Becky must be lying. This means that Ducky is telling the truth. Therefore Cindy is the only possibility.

Look under Logic in the Index for similar problems.

S2000-2

The solution to part **(a)** is $2000 = 5555_7$.

(b) $1885 = 1111_{12}$. Let the number be $N = aaaa_b = aM_b$ where

$$M_b = (1 + b + b^2 + b^3).$$

Then since $1 \le a \le b - 1$,

$$M_b \le N \le (b-1)M_b < b^4.$$

Since $N < 2000$, $M_b < 2000$, $b \le 12$ and since $1111_{12} = 1885$, we have $N \ge 1885$ whence $b^4 > 1885$ so that $b \ge 7$. The values of M_b for $b = 7$, 8, 9, 10, and 11 are 400, 585, 820, 1111, 1464 and no multiples of these lie between 1885 and 2000.

Look under Number Theory in the Index for similar problems.

S2000-3

It converges. Let

$$S_n = \sqrt{1 + \sqrt{2 + \sqrt{3 + \ldots + \sqrt{n}}}}.$$

Since $n + \sqrt{n+1} > n$, $S_{n+1} \ge S_n$ and therefore S_n are increasing. We show the $S_n < 2$ for all n and thus we have convergence. For $x \ge 6$, we have

$$x + \sqrt{2x} < 2x - 2.$$

Thus we see that

$$S_n < \sqrt{1 + \sqrt{2 + \sqrt{3 + \cdots + \sqrt{n-2+\sqrt{n-1+\sqrt{2n-2}}}}}}$$

$$< \sqrt{1 + \sqrt{2 + \sqrt{3 + \cdots + \sqrt{n-2+\sqrt{2n-4}}}}}$$

$$< \sqrt{1 + \sqrt{2 + \sqrt{3 + \cdots + \sqrt{n-3+\sqrt{2n-6}}}}}$$

$$< \cdots$$

$$< \sqrt{1 + \sqrt{2 + \sqrt{3 + \sqrt{4 + \sqrt{5 + \sqrt{10}}}}}}$$

$$< \sqrt{1 + \sqrt{2 + \sqrt{3 + \sqrt{4 + \sqrt{9}}}}}$$

$$< \sqrt{1 + \sqrt{2 + \sqrt{3 + 3}}}$$

$$< \sqrt{1 + \sqrt{5}} < 2.$$

The actual value is $1.75793\ldots$.

Look under Sequences in the Index for similar problems.

S2000-4

The order of an element divides the order of G, and can therefore be 1, 3, 5, or 15. For these k, let n_k be the number of elements of order k. Clearly $n_1 = 1$. We'll show that $n_{15} > 0$ (an element x of order 15 exists) and so $G = <x>$ is cyclic. Suppose G has A subgroup(s) of order 3 and B subgroup(s) of order 5. Since 3 and 5 are prime, by Sylow's theorem, $A \equiv 1 \pmod 3$, $B \equiv 1 \pmod 5$, and both A and B divide 15. Thus $A = B = 1$ and $n_3 = 2$, $n_5 = 4$. Since $n_1 + n_3 + n_5 + n_{15} = 15$, $n_{15} = 15 - 1 - 2 - 4 = 8 > 0$.

Look under Group Theory in the Index for similar problems.

S2000-5

Since

$$\frac{1}{\sqrt{5}}\tau^k - \frac{1}{\sqrt{5}}(1-\tau)^k$$

is an integer for $k \geq 1$, and $|1 - \tau| < 1$, we have

$$F_k = \frac{1}{\sqrt{5}}\tau^k - \frac{1}{\sqrt{5}}(1-\tau)^k$$

so

$$\sum_{k=0}^{\infty} \frac{F_k}{k!} = \frac{1}{\sqrt{5}}(e^{1/2}e^{\sqrt{5}/2} - e^{1/2}e^{-\sqrt{5}/2}) = 2\sqrt{e/5}\sinh\left(\frac{\sqrt{5}}{2}\right),$$

$$\sum_{k=0}^{\infty} \frac{(-1)^k F_k}{k!} = -\frac{2}{\sqrt{5e}}\sinh\left(\frac{\sqrt{5}}{2}\right)$$

and

$$\left(\sum_{k=0}^{\infty} \frac{F_k}{k!}\right)\left(\sum_{k=0}^{\infty} \frac{(-1)^k F_k}{k!}\right) = \frac{-4}{5}\sinh^2\left(\frac{\sqrt{5}}{2}\right) = \frac{2}{5}(1 - \cosh\sqrt{5})$$

by the double angle formula $\cosh 2x = 2\sinh^2 x + 1$.

Look under Fibonnacci Sequences or Infinite Series in the Index for similar problems.

S2000-6

The desired ratio is $\rho = K/\pi R^2$, where K denotes the area of the triangle. We use the law of sines, $\sin C = c/2R$ and the area formula $K = \frac{1}{2}ab\sin C$ to write

$$\rho = \frac{(a/R)(b/R)\sin C}{2\pi} = \frac{2}{\pi}\sin A\sin B\sin C$$

$$= \frac{2}{\pi}\sin 2B\sin B\sin 3B$$

$$= \frac{4}{\pi}\cos B\sin^3 B(4\cos^2 B - 1)$$

$$= \frac{4}{\pi}x(1-x^2)^{3/2}(4x^2 - 1)$$

where $x = \cos B \in (1/2, 1)$. Thus we seek the maximum of

$$\rho = \frac{4}{\pi}f(x), \quad f(x) = (1-x^2)^{3/2}(4x^3 - x) \quad \text{on } (1/2, 1).$$

$$f'(x) = (1 - x^2)^{3/2}(12x^2 - 1) + (4x^3 - x)(3/2)(1 - x^2)^{1/2}(-2x)$$
$$= (1 - x^2)^{1/2}\left[(1 - x^2)(12x^2 - 1) + (4x^3 - x)(-3x)\right]$$
$$= (1 - x^2)^{1/2}(16x^2 - 1 - 24x^4)$$

so

$$16x^2 - 1 - 24x^4 = 0 \quad \text{or} \quad x^2 = \frac{4 + \sqrt{10}}{12},$$

the other root being less than $1/2$. Thus the answer is

$$\frac{4}{\pi}g(-1g^2)^{3/2}(4g^2 - 1).$$

This can be written as

$$\frac{4}{\pi}\left(1 - \left(\frac{1}{3} + \frac{1}{12}\sqrt{10}\right)\right)^{3/2}\left(4\left(\frac{1}{3} + \frac{1}{12}\sqrt{10}\right) - 1\right)\sqrt{\frac{1}{3} + \frac{1}{12}\sqrt{10}}$$

$$= \frac{7\sqrt{10} - 2}{108\pi}\sqrt{22 + 4\sqrt{10}}$$

$$= \frac{(7\sqrt{10} - 2)(\sqrt{20} + \sqrt{2})}{108\pi}$$

$$= \frac{34\sqrt{2} + 5\sqrt{5}}{54\pi}.$$

Look under Geometry or Max/Min Problems in the Index for similar problems.

S2000-7

Consider $x = \sqrt{2}^{\sqrt{2}}$. If x is rational, it is a counterexample. Otherwise, $y = x^{\sqrt{2}}$ is a counterexample since $y = 2$ is rational.

Look under Number Theory in the Index for similar problems.

S2000-8

If $Df = f'$, the characteristic polynomial factors as

$$(D^2 + 1)(D - 1)(D^2 + D + 1)f = 0,$$

which yields five linearly independent solutions, which can be combined under the Principle of superposition to yield the general solution

$$f(x) = \sum c_i e^{r_i x},$$

where r_i is a root of the characteristic polynomial $(\pm i, 1, \omega, \overline{\omega})$, with $\omega = (-1 + \sqrt{3}i)/2$. Since $\lim f(x) = 0$, $Re(r_i) < 0$ and $f(x) = Ae^{i\omega} + Be^{i\overline{\omega}}$. Since $f(0) = 0$, $B = -A$ and

$$f(x) = Ae^{-x/2}\left[e^{ix\sqrt{3}/2} - e^{-ix\sqrt{3}/2}\right]$$

$$= Ce^{-x/2}\left(\sin\frac{\sqrt{3}}{2}x\right)$$

for some constant C.

Look under Differential Equations in the Index for similar problems.

Index by Problem Type

Algebraic Structures
1971-1, 1971-7

Analytic Geometry
1967-2, 1976-5, 1978-2, 1979-6, 1981-6, 1988-6, 1990-6, 1991-1, 1991-7, 1994-6, 1996-4, 1997-6, 1998-4

Arclength
1992-1

Binomial Coefficients
1999-6

Complex Numbers
1967-5, 1968-7, 1986-2, 1986-3, 1995-1, 1997-3

Derangements
1987-4, 1998-6

Differentiation
1966-4, 1971-3, 1976-3, 1982-4, 1996-8

Differential Equations
1969-4, 1979-3, 1980-3, 1988-1, 2000-8

Diophantine Equations

1966-1, 1968-4, 1983-6, 1992-8, 1996-7

Enumeration

1966-5, 1966-8, 1967-4, 1968-8, 1971-4, 1971-6, 1975-3, 1982-2, 1984-2,
1984-7, 1985-4, 1986-4, 1987-2, 1990-2, 1991-2, 1991-3, 1991-7, 1992-3,
1997-1

Field and Ring Theory

1973-5

Fibonnacci Sequences

1983-2, 1998-4, 2000-5

Finite Sums

1978-3, 1980-7, 1995-2, 1997-2

Fundamental Theorem of Calculus

1988-3

Geometry

1966-2, 1967-6, 1967-8, 1968-2, 1969-5, 1970-3, 1970-4, 1975-2, 1978-5,
1981-3,1982-3, 1983-3, 1987-5, 1988-5, 1988-6, 1989-3, 1990-4, 1991-3,
1993-8, 1994-4, 1995-7, 1996-2, 1996-5, 1997-7, 1998-1, 1999-2, 2000-6

Group Theory

1972-7, 1974-2, 1985-3, 1988-2, 1992-4, 1993-2, 2000-4

Inequalities

1980-1, 1994-2, 1994-3, 1995-5, 1999-3

Infinite Series

1967-7, 1968-5, 1968-6, 1972-4, 1974-6, 1975-1, 1983-4, 1987-3, 1999-7,
2000-5

Integration

1971-3, 1973-4, 1977-2, 1980-6, 1987-6, 1988-5, 1989-6, 1992-1, 1992-5, 1995-5, 1998-5

Limit Evaluation

1966-6, 1968-5, 1970-1, 1971-5, 1973-1, 1973-4, 1976-6, 1983-1, 1984-6, 1985-5, 1990-5, 1991-6, 1992-7, 1993-5, 1994-1, 1996-6

Logic

1989-1, 1989-7, 1993-7, 1999-1, 2000-1

Matrix Algebra

1967-1, 1972-1, 1972-2, 1973-7, 1977-3, 1980-4, 1986-2, 1989-6, 1992-6, 1993-3, 1998-7, 1999-8

Max/Min Problems

1973-2, 1981-5, 1983-3, 2000-6

Miscellaneous

1966-7, 1974-3, 1979-1, 1979-5

Multivariate Calculus

1968-3, 1974-4, 1975-5, 1986-6

Number Theory

1966-3, 1969-2, 1969-3, 1970-5, 1971-2, 1973-8, 1975-4, 1976-2, 1977-5, 1978-1, 1980-2, 1981-1, 1982-6, 1983-7, 1984-1, 1984-3, 1984-4, 1985-2, 1985-6, 1987-1, 1988-4, 1990-1, 1991-4, 1991-5, 1993-6, 1994-5, 1995-3, 1996-1, 1996-3, 1998-2, 1998-3, 2000-2, 2000-7

Permutations

1972-5, 1993-4

Probability

1973-9, 1978-6, 1981-4, 1982-6, 1983-2, 1984-5, 1985-7, 1987-4, 1988-7, 1989-4, 1992-2, 1996-6, 1997-1, 1999-5

Polar Coordinates

Polynomials

Real-Valued Functions

Riemann Sums

Sequences

Solving Equations

Systems of Equations

Statistics

Synthetic Geometry

Taylor Series

Trigonometry

Volumes